SCIENTIFIC PERSPECTIVES OF TEA PLANT HORTICULTURE AND PRODUCTIVITY

SCIENTIFIC PERSPECTIVES OF TEA PLANT HORTICULTURE AND PRODUCTIVITY

L. MANIVEL, PHD (UC DAVIS, CA., USA)

Expert on Tea and Horticultural Crops, India
Former Plant Physiologist, Tocklai Tea Research Institute and UPASI Tea Research Foundation, India

ELSEVIER

ACADEMIC PRESS

An imprint of Elsevier

Academic Press is an imprint of Elsevier
125 London Wall, London EC2Y 5AS, United Kingdom
525 B Street, Suite 1650, San Diego, CA 92101, United States
50 Hampshire Street, 5th Floor, Cambridge, MA 02139, United States
The Boulevard, Langford Lane, Kidlington, Oxford OX5 1GB, United Kingdom

Library of Congress Cataloging-in-Publication Data
A catalog record for this book is available from the Library of Congress

British Library Cataloguing-in-Publication Data
A catalogue record for this book is available from the British Library

ISBN: 978-0-12-823444-0

Publisher: Charlotte Cockle
Acquisitions Editor: Charlotte Cockle
Editorial Project Manager: Isabella C. Silva
Production Project Manager: Swapna Srinivasan
Cover Designer: Mark Rogers

Typeset by TNQ Technologies

Contents

Foreword

Dr. Laxman Manivel has written a book titled *Scientific Perspectives of Tea Plant Horticulture and Productivity*. This encompasses, in general, the overall physiology, agronomy, and chemistry of the well-known crop plant known as tea. Unlike the coffee plant, the tea plant, growing from the seedlings/ cuttings to the final manufacture of tea, namely, the tea dust that we use as a beverage, involves extensive agronomic and technological planning and programming. With commendable efforts based on his training at the University of California, Davis, his dedicated work at Tocklai, a premier research center for tea research and technology in India, for over 40 years, and including his advisory of tea in neighboring Sri Lanka, Dr. Manivel is placed as a remarkable contributor throughout his book.

The book written by Dr. Manivel is comprehensive, not lacking even in a small feature. The author is highly qualified in horticulture, which includes, among other crops, tea (*Camelia sinensis*), and a lifetime of research in tea growing in the world-famous state of Assam. He initiated research using $C^{14}O_2$ on the photosynthesis of this field crop. He has also initiated and established a laboratory for plant physiological research at Tocklai.

He has studied tea growing from the planting to harvest and manufacturing of tea in the final beverage form for over 20 years, experimenting for his research with growth substances and organic fertilizers, among many other components.

I have known Dr. Manivel for over 30 years when he first came to my department at the Bhabha Atomic Research Centre, Bombay, for analyzing leaf samples. He was also keenly interested in knowing the use of radioactive nutrient elements, especially in their absorption and distribution in tea plants.

It is my great privilege and pleasure to write this small Foreword, which only mentions a few parts among the large information diligently assembled from his work and from the literature in a well-organized presentation. This volume will go a long way to give full information for the scientists and growers.

The author is highly qualified, worked in the state department of agriculture and the agricultural Universities in Coimbatore and Madurai in Tamil Nadu. His expertise gained at Davis on the photosynthetic patterns in grapes has been usefully employed in his main research on tea in India and Sri Lanka, which he was invited for advice and consultation.

Seshadri Kannan
Retired Head, Plant Nutrition section at BARC
Recipient of MSU Distinguished Alumnus (1987) and MSU Honored International Award in Recognition of leadership, and contributions to advancing friendship, knowledge, and understanding around the world (1993)

Preface

Tea is an evolved beverage used for pleasure and health. A good cup of tea in the morning and evening cheers up and recharges. Its origin, development, and spread have been amply documented. This is one of several agri-horticultural plantation crops with advanced field and processing practices, with marketing of the produce developed and streamlined by the pioneers, which has made tea one of the premier world beverage crops.

Taxonomically, tea (*Camellia sinensis*, Theaceae) is a tree by habit, and it has been managed as a bush to facilitate harvesting by workers, especially women. The plant part harvested periodically is also unique—the emerged succulent flush of two leaves and the bud. It is pertinent to note that the flush harvested in the south is of the three-plus bud, which is more fibrous and less succulent instead of two-plus bud in the north. This is used for making the sought-after commercial tea.

The field packages like pruning, training the bush, architecture, moderating the ambient climate for economic growth and productivity of the bushes through judicious planting of legume trees, design of inputs, processing, marketing, and providing the infrastructure are unique, and hats off to the brave, intelligent, and hardworking pioneers!

Working with tea is a pleasure and privilege. We are working with an organized living creature, disciplined, sincere, dedicated, hardworking, toiling day and night, and compensated with rewards and incentives. Although forestland has been utilized for tea plantations, the industry is conscious of the environment protection by maintaining three-tier vegetation thoughtfully planned, designed, and executed. Despite the isolation and adversarial conditions, working in tea plantations leads to pleasure with longer life into late senescence. Considering the global change of climate and pollution due to industrialization and real estate, plantations are fairly free, despite the fact the soil and environment affected by hard chemicals imported and used indiscriminately, which ruined the soil/plant biospheres in the last two decades, needs to be rectified and restored on priority.

While I was traveling from Kolkata to Jorhat for joining Tocklai on February 01, 1976, a planter friend, Tiwari of Octavius Steel, pointed down and showed the pruned fields of tea in the plains. My eyes were opened to confirm there were tea plantations in the plains. Because I was familiar with the plantations in Travancore, Kerala, Nilgiris, Anamallais, and Tamil Nadu, which grow in rolling hills, evergreen, and with flushing throughout the year, I realized the variations of field situations include growing and flushing behavior of bushes, crop distribution, inputs, harvesting, manufacturing, and types of made tea between North India and Southern India including Sri Lanka.

Ethnically, the plantation workers in southern India, Sri Lanka, Malaysia, Mauritius, Indonesia, Parts of South Africa are Tamil, while the northern plantations possess the work force of Central and North India, Bihar, Orissa, Madhya Pradesh, and Andhra. These ethnically differing worker populations of about 25 lakhs merged and integrated with local population to an extent in five generations, speak a mixed common language, lost their roots, and have emerged as the formidable, dependable, disciplined, skilled work force awaiting the benefits of welfare measures for rehabilitation, uplift, settlement, consensus in collaboration with the concerned state governments to join the mainstream and continue to enjoy the rights of a citizen of independent India. Health and efficiency of the workforce play an important role on growth, productivity, quality, sustainability, and profit margin of the plantations. The present conditions of the tea industry and its potential, based on the unique, complex, historical background, social impacts, and so forth, deserve due consideration bearing in mind that the productivity and quality are influenced by many factors such as climate, soil terrain, cultivars, processing and manufacturing, and package of practices, in addition to the efficiency of work force. These are to be taken into account by the administration/authorities, and all the stakeholders of the industry, before implementing the ensuing package for the revival of the tea plantations of the nation.

The present condition of stagnating crop productivity with deteriorated quality, coupled with non-Remunerative price realizations, are due to the overexploitation of bushes that impact the soil, plant biosphere, weakening the bushes predisposed to pests and diseases, and escalating the COP. Consequently the bushes are going through chronic stress due to inbuilt toxins, weakened defoliation, die back, and mortality, causing excessive vacancies of both tea bushes and shade trees. Remedial measures for restoring the health and productivity lies on the reclamation of soil—plant biospheres, withdrawing all the hard chemical uses, moderating inputs, integrating the INM and IPM, encouraging, bioproducts in an ecofriendly way, adopting natural farming, GAP optimizing the soil parameters, ensuring timely effective harvest of the crop, and processing and marketing the crop effectively. The turnaround is expected in a year, which is a win-win situation.

Another important aspect is the uprooting age for tea plantations. Because tea is a perennial crop, it has no age limit provided the optimum conditions for growth are provided. Any deficiency or sickness debilitates the bushes, and weaker bushes become sick and die. It is not natural but happens from mismanagement. The cambium tissue in the vascular bundle generates new growth every year, forming annual rings in perpetuity. The marginal surplus lands can be uprooted, reclaimed, and replaced with inter-planted/consolidated with suitable horticultural, spice crops, afforestation, and so forth, with advantage. This will be an effective alternative for economical, judicious discretion for effective better utilization of surplus and marginal lands in the gardens. Forage crops with dairying can also be thought of as part of diversification and value addition.

Divine guidance and encouragement from faculty members, along with the timely help and encouragement from family, well-wishers, and close friends, have resulted in this book: a small contribution to the scientific society of the world. I thank them, and I am grateful forever.

I dedicate this book for the following faculty members at UCD who taught and trained me:

Robert J. Weaver, Mentor, Phytohormones, Grapes
Julian Crane, Graduate Adviser, Fig, Pistachio
Eric C. Conn, Plant Biochemistry
Robert W. Glocks, Vice Chancellor, Finance
Harold P. Olmo, Plant Breeder, Grapes
Albert J. Winkler, Emeritus Scientist, Viticulture

L. Manivel
Coimbatore
February 11, 2020

CHAPTER ONE

Botany origin and spread of tea cultivars

Contents

Camellia sinensis (L) O.Kuntz is the botanical name of cultivated tea, having three recognized subspecies, such as Assam *Camellia sinensis* var. *assamica*; *Camellia sinensis* sub. sp., *lasiocalyx, Cambod, C. Camellia sinensis* var. *sinensis*, with origin in different places like Assam (Saikowa, Brahmaputra valley), Cambod (Iravathy River valley, Indochina) and the Yunnan valley (Yiantsikiang river valley of China), the oldest known and cultivated by China belonging to Theaceae.

Among the three cultivated species, Cambod is the most vigorous and tall growing, with high potential of yield, but it suffers in quality. Assam is the next-most vigorous, with broad succulent bulliform leaves contributing for the strong liquor and aroma sought after for blending and packaging. China cultivates the bushy type with slender stem and small, semierect orientation of leaves (often with a purple tinge and prominent serration), known for the rich flavor like Darjeeling premier tea. Tea is a C-3 plant occupying the bottom of the ladder of evolution, with a low rate of photosynthesis and higher loss of carbon due to photorespiration, compared to the C-4 plants, which are more evolved. It is a tree but trained as a bush/shrub through periodical pruning to facilitate easy harvesting of immature shoots (flush), which are used for the manufacture of the commercial tea.

The British East India Company sent explorers from Calcutta with army personnel who brought China variety seeds from China and introduced in the Botanic Garden, Howrah, Dibrugarh in Assam, Wyanad and Nilgiris in South India, and Himachal in north India, successfully establishing the

Scientific Perspectives of Tea Plant Horticulture and Productivity
ISBN: 978-0-12-823444-0
https://doi.org/10.1016/B978-0-12-823444-0.00001-7

China tea. The first consignment made to the UK was well received, and the East India Company with the Army helped explore it in Assam, where they picked up the broad-leafed (Betjan) Assam seed jat, by Major Robert Bruce accompanied by Assamese local planter Diwan in 1816. Their search continued, and they picked up the elite plant in 1888. After selection, propagation, and evaluation in field studies during the end of the 19th century, the elite plants of three types of tea had been procured and introduced for evaluation, comparative examination, and the later release of Tocklai vegetative cultivar (TV) clones.

Through the introduction of tea and selection activities initiated in 1911, at the Kolkata Botanic Garden, the present Tea Research Station was established in Tocklai, Jorhat, Assam in 1930 with scientist (Dr. William Wight) as botanist, initiating selection and breeding of new varieties of tea and propagating tea vegetatives. New species like *Irrawadiensis* and *Camellia sasanqua* were added to the germ plasm.

Breeding and selection of new varieties were intensified. Triploids and polyploids were produced and new clones with gene map and traits were documented through the botany Tocklai, under the leadership of the senior botanist and plant physiologist, until 1962. Proven progenies and cultivars were distributed to Sri Lanka, Kenya, South Africa, and later to other countries like Japan, Russia, Australia, Argentina, and Turkey through this Institute. (For more details on the distribution and spread of tea, please refer to Percival Griffiths, 1951; Barua, 1989, Baruah, 2006, 2019a,b; Carr, 2018).

The areas where tea was cultivated steadily expanded; consequently, demand increased for planting materials and new improved varieties of tea. Breeding, pure-line selection, and establishment of seed orchards for biclonal stocks were intensified, contributing new elite cultivars in the 1980s and meeting the expansion programs of the industry. TV1 was the first elite clone picked up and released from Tocklai in 1954. A good number of TV clones, biclonal stocks, and pure-line selections were released and added to the germ plasm. Quite a few garden clones were also released by pure-line selection and introduced for regional acceptancy, productivity, as well as quality (a few are listed here: S3A1, S3A3, T3E3 Jorhat, Assam type, P126, NH 431, R92, AV2 Chinatype, TeenAli17, TRI, 2024, 2025, and 2043 Cambod type). Thus, these decades (1970s and 1980s) were the most productive period for Pioneer, Tocklai Experimental Station, Jorhat, and Assam, supporting the tea industry and boosting the area, productivity, quality, and value realization (Tocklai Conf. Proceedings, 1981; Barua, 1989; Baruah, 2019a,b). Though many clones were released later for different aspects/purposes, the first clone, TV1, is still considered number one for productivity, quality, stress tolerance, and for being a good blending clone despite its few setbacks such as brown, fibrous-made tea, and susceptibility to red spider mite. It is a potent mother-bearer cultivar, evolving biclonal stocks, as well as a mother plant for future breeding programs.

The positive effects of tap root in seedlings in terms of establishment and stress tolerance were well recognized. Similarly, the hybrid vigor of combining elite clones in the natural environment was appreciated by the industry. This led to the establishment of seed baris combining various proven elite productive stress-tolerant varieties. The progenies were evaluated in regional trials, and the best biclonal stocks were released to the industry. Cambod, being known for vigor, were used as pollinators (male), while the productivity, quality, and stress tolerance of the elite Assam/China cultivars were used as bearer plants (female), exploiting both the hybrid vigor as well as the beneficial traits.

S3A3 x TV1 = TS 491

TV1 x TV18 = TS 462

TV1 x TV20 = TS 520

TV20xH23/14 = TS 581

1.1 Characteristics of the species (a list of Assam, Cambod, and Chinary tea bushes TV3, TV 7, TV 9, AV2, and P126)

Assam: Genetic makeup $2n = 30$, broad bullish, light green laminae, horizontal leaves with lanceolate tip, stem brownish, succulent, very prominent shoot tip, and fairly pubescent. Bushes go dormant from mid-December—mid-February. Early riser and early sleeper. Follows phasic growth in flushing up to six to seven phases. Native of Assam: nanoamino acid and micronutrients foliar.

Cambod: Genetic makeup $2n = 30$, with medium sized semierect leaves, with prominent serrated leaf margin and characteristic pink petiole, stem medium, with a thick, ashy color. Slightly waxy laminae contributing to stress tolerance, with prominent leaf tip. Completes six to seven phases of growth with interflush dormancy. Early riser and late sleeper. Endogenous level of triacontanol (C30-ethoxy fatty acid alcohol) being high, the variety does not respond to the foliar of triacontanol even with zinc.

China: Genetic makeup $2n = 30$, with a smaller frame, thinner sticks, and smaller semierect waxy leaves. Late riser, and late sleeper contributing to tail end crop, low productive with quality of rich flavor due to carotenoids, sclerides, synthesizing salicylic acid (mevalonic pathway), responding to manganese-rich soil and foliar. Goes through longer interflush dormancy completing about six phases of growth; handicapped due to smaller shoots; low productivity with characteristic silvery pubescence contributing for the unique flavor and light liquor. Responds well for foliar of all growth regulators including nano biostimulants with micronutrients. The tea plants flower in spring (April) and the fruits (capsules) mature in the fall (November). The seeds drop from the bursting capsules and are collected from the ground periodically, graded by floating test and stored in moist charcoal/sand media. The seed material is quality controlled and certified for purity through periodical inspections. After a reasonable period of exploitation of seed production, the bushes can be top worked with required clones for seed production as a bearer, while the *Cambod* are used as pollinators. Cleft grafting is done for conversion.

1.2 Germ plasm preservation of tea

Breeding for evolution of improved cultivars while meeting the changed situations in the future is a continuous process. Preservation of the wild species and elite cultivars in germ plasm is essential for the future of the industry. A germplasm of tea has been maintained in Tocklai, TRI, and UPASI TRF both for Northern India and Southern India in addition to the preservation in the Bureau of Plant Introduction in Delhi, IARI.

1.3 Statistics

History of Indian Tea.

Year	No. of tea estates †	Area under tea (Hectares)	Production (Tons)	Yield per hectare kg.	Exports !			Retention (Tons)	Sold in indian auctions*		Sold in london auctions □	
					Quantity in tons	Value in thousand rupees	Value per kg. Rs.		Tons	Avg. per kg. Rs.	Tons	Avg. per kg. Pence
1885	—	114,900	32,444	282	30,897	—	—	1547	—	—	—	—
1895	—	168,234	65,049	387	61,016	7,50,50	1.23	4033	—	—	—	—
1900	—	211,443	89,567	424	87,226	9,59,49	1.10	2341	24,569	0.70	—	—
1905	5696	213,675	100,567	471	98,325	8,94,76	0.91	2242	27,719	0.75	53,249	6.9
1910	4402	228,062	119,569	524	116,319	12,56,25	1.08	3250	34,470	0.96	68,601	7.9
1915	4437	256,951	168,829	657	154,418	20,07,43	1.30	14,410	47,706	1.22	90,450	10.0
1920	5054	284,922	156,645	550	130,419	12,25,94	0.94	26,226	37,474	0.68	49,783	13.4
1925	4338	294,474	164,886	560	153,003	28,15,26	1.84	11,883	39,688	1.78	52,276	15.7
1930	4743	325,057	177,391	546	164,243	23,97,95	1.46	13,148	42,043	1.22	—	13.4
1935	5134	336,572	178,912	532	146,926	19,82,41	1.35	31,986	48,263	0.99	115,485	11.7
1940	6564	337,296	210,415	624	160,664	27,89,92	1.74	49,751	45,274	0.83	□	□
1945	6685	308,922	229,038	741	167,638	36,28,74	2.16	61,400	48,252	2.04	□	□
1950	6731	315,656	278,212	881	200,780	80,42,14	4.01	77,432	113,645	4.23	4.56	38.2
1955	6840	320,238	307,704	961	166,708	113,61,32	6.82	140,996	166,325	4.35	83,905	57.1
1960	9499	330,738	321,077	971	193,063	119,98,83	6.21	128,014	173,791	5.28	76,795	52.1
1965	10,823	341,762	366,374	1072	199,365	114,97,47	5.77	167,009	221,084	5.35	86,031	47.0
1970	12,015	354,133	418,517	1182	202,335	149,80,33	7.40	216,182	241,311	6.47	32,669	46.3
1975	13,166	363,303	487,137	1341	218,480	246,02,13	11.26	268,657	286,593	10.59	40,806	63.3

(Continued)

History of Indian Tea.—cont'd

					Exports !				Sold in indian auctions*		Sold in london auctions □	
Year	No. of tea estates †	Area under tea (Hectares)	Production (Tons)	Yield per hectare kg.	Quantity in tons	Value in thousand rupees	Value per kg. Rs.	Retention (Tons)	Tons	Avg. per kg. Rs.	Tons	Avg. per kg. Pence
1980	13,382	381,086	569,172	1494	224,780	432,54,61	19.24	344,392	306,959	13.60	29,598	104.1
1985	13,537	399,966	656,162	1641	214,937	703,59,04	32.73	441,225	505,291	23.34	10,053	198.5
1986	13,546	407,647	620,803	1523	204,292	590,01,31	28.88	416,511	468,405	23.44	24,108	141.9
1987	13,564	411,335	665,251	1617	202,753	646,55,73	31.89	462.498	472,474	24.66	12,082	122.5
1988	13,853	414,347	700,014	1689	201,747	619,00,34	30.68	498,267	497,476	24.36	8826	125.5
1989	13,856	414,953	688,105	1658	212,662	848,97,80	39.92	475,443	477,386	36.62	7399	144.8
1990	13,861	416,269	720,338	1730	210,024	1,113,35,10	53.01	510,314	482,251	43.23	9141	142.3
1991	13,873	420,470	754,192	1794	202,918	1,134,55,33	55.91	551,274	501,585	40.31	6033	132.0
1992	13,918	420,289	732,322	1742	174,962	995,33,07	56.89	557,360	448,061	38.88	6666	116.6
1993	13,936	418,363	760,826	1819	175,318	1,161,26,36	66.24	585,508	441,718	48.93	5145	130.96
1994	35,116	425,966	752,895	1768	150,691	989,14,69	65.64	602,204	428,303	40.61	5195	115.23
1995	37,319	427,065	756,016	1770	167,996	1,208,01,56	71.91	588,080	428,460	47.99	3075	126.95
1996	38,705	431,204	780,140	1809	161,696	1,246,87,19	77.11	618,444	443,116	48.77	2491	126.65
1997	38,707	434,294	810,031	1865	202,995	1,774,77,81	87.43	607,036	459,026	66.89	1657	148.31
1998	88,010	474,027	874,108	1844	210,338	2,309,43,60	109.80	663,767	442,359	76.43	471	205.38
1999	98,839	490,200	825,935	1685	191,719	1,965,86,84	102.54	632,216	480,768	72.79	□	□
2000	112,010	504,366	846,922	1679	206,816	1,898,61,19	91.80	640,106	507,155	61.71	□	□
2001	115,264	509,806	853,923	1675	182,588	1,682,11,15	92.13	671,335	463,688	61.66	□	□
2002	115,264	515,832	838,474	1575	201,002	1,753,38,98	87.23	637,472	456,549	55.95	□	□
2003	129,027	519,598	878,129	1690	173,684	1,590,21,28	91.56	683,371	473,370	56.28	□	□
2004	129,027	521,403	892,965	1713	197,668	1,841,14,16	93.14	695,297	455,948	65.50	□	□
2005	140,712	555,611	945,974	1703	199,050	1,830,97,86	91.99	746,924	496,569	58.67	□	□
2006	143,217	567,020	981,805	1732	218,734	2,006,52,60	91.73	763,071	484,552	66.60	□	□
2007	159,190	578,458	986,427	1705	178,754	18,101,095	101.26	807,673	502,763	67.97	□	□

					Exports !				Sold in indian auctions*		Sold in london auctions □	
Year	No. of tea estates †	Area under tea (Hectares)	Production (Tons)	Yield per hectare kg.	Quantity in tons	Value in thousand rupees	Value per kg. Rs.	Retention (Tons)	Tons	Avg. per kg. Rs.	Tons	Avg. per kg. Pence
2008	—	560,480	980,818	1750	203,117	23,929,136	117.81	777,701	530,020	87.78	□	□
2009	—	560,624	978,999	1746	197,903	27,858,465	140.77	781,121	501,479	106.70	□	□
2010	—	579,353	966,403	1724	222,019	30,583,081	137.75	744,404	512,301	106.32	□	□
2011	—	563,979	1,115,720	1926	215,412	32,918,710	152.83	917,248	534,310	106.07	□	□
2012	—	563,979	1,126,330	1997	208,260	37,507,600	180.10	938,690	548,017	125.30	□	□
2013	—	566,663	1,200,410	2128	219,060	43,552,300	198.81	981,350	572,683	132.23	□	□
2014	—	566,663	1,207,310	2131	207,440	40,540,400	195.43	999,870	568,752	130.91	□	□
2015	—	566,663	1,208,660	2133	228,660	43,553,200	190.47	980,000	589,725	128.60	□	□
2016	—	577,477	1,267,360	2195	222,450	44,681,100	200.86	1,044,910	542,127	135.93	□	□
2017	—	599,684	1,321,760	2240	251,911	49,875,900	197.99	1,069,849	623,719	134.81	□	□
2018	211,794	636,557	1,338,630	2102	256,060	53,353,300	208.36	1,082,570	620,472	140.80	□	□
2019	—	—	1,389,700	—	248,290	56,106,500	225.97	1,141,410	619,639	142.15	□	□

† Figures include small growers and exclude the inoperative estates of the district.
— Denotes not available.
□ London Auctions stopped from 25th August 1939 due to World War and resumed from 16th April 1950. London Auction Center was finally closed in June 1998.
!Exports upto 1950 relate to seasonal year, thereafter calendar year.
*Indian Auctions upto 1963 relate to seasonal year, thereafter calendar year.
Area under tea.
Source: Tea Board except 2008 and 2009 "International Tea Committee".

Area under tea **India**

(In hectares)

	2005	2006	2007	2008	2009	2010	2011	2012	2013	2014	2015	2018
North India												
Darrang	41,388	41,677	41,710	–	–	–	–	–	–	–	–	–
Goalpara	4215	4451	4451	–	–	–	–	–	–	–	–	–
Kamrup	3787	3782	3953	–	–	–	–	–	–	–	–	–
Lakhimpur	6301	6724	6756	–	–	–	–	–	–	–	–	–
Dibrugarh	114,435	120,489	122,514	–	–	–	–	–	–	–	–	–
Nowgong	8605	8709	8758	–	–	–	–	–	–	–	–	–
Sibsagar	83,971	88,008	94,611	–	–	–	–	–	–	–	–	–
Assam Valley	268,608	280,017	289,006	–	–	–	285,830	–	270,920	–	307,080	–
Cachar	31,894	31,805	32,313	–	–	–	36,380	–	33,480	–	–	–
Assam†	300,502	311,822	321,319	–	–	–	322,210	–	304,400	–	–	337,690
Darjeeling	17,539	17,542	17,818	–	–	–	17,820	–	17,820	–	–	–
Terai	24,313	24,340	24,359	–	–	–	24,360	–	49,700	–	–	–
Dooars	72,673	72,906	72,918	–	–	–	72,920	–	72,920	–	–	–
West Bengal	114,525	114,788	115,095	–	–	–	115,100		140,440	–	140,440	148,122
Tripura	8710	8710	8962	–	–	–	–	–	–	–	–	–
Others!!	12,051	12,051	13,342	–	–	–	–	–	12,290	–	12,290	49,817
Total N.I.	435,788	447,371	458,718	447,140	447,106	446,738	459,610	457,130	457,130	459,810	459,810	535,629
South India												
Tamilnadu	80,939	81,276	80,462	–	–	–	80,460	–	69,620	–	69,620	62,885
Karnataka	2112	2137	2141	–	–	–	2140	–	2220	–	2220	2172
Kerala	36,772	36,236	37,137	–	–	–	37,140	–	35,010	–	35,010	35,871
Total S.I.	119,823	119,649	119,740	113,340	113,518	113,871	119,740	106,850	106,850	106,850	106,850	100,928
Grand total	555,611	567,020	578,458	560,480	560,624	560,609	579,350	563,980	563,980	566,660	566,660	636,557

† Includes North Cachar & KarbiAnglong.
!! Including Bihar, Uttaranchal, Himachal Pradesh, Sikkim, Meghalaya, Manipur, Orrisa, Arunachal Pradesh and Nagaland.
– Denotes not available.
Source: Tea Board, except 2008 to 2010 and 2012, 2014. "International Tea Committee."

Area under tea **World**

(In hectares)

	2008	2009	2010	2011	2012	2013	2014	2015	2016	2017	2018
India	560,480	560,624	579,353	563,979	563,979	566,663	566,663	566,663	577,477	599,684	636,557
Sri Lanka	188,323	188,175	188,007	187,860	187,000	187,000	187,935	188,000	202,839	202,540	202,540
Indonesia	139,416	124,056	122,796	122,764	122,206	122,494	121,034	119,361	117,268	116,500	115,300
Bangladesh	54,106	54,700	54,900	54,400	54,500	55,000	56,000	56,846	59,000	59,000	59,300
Nepal	16,718	17,127	17,451	18,149	19,036	20,120	26,165	26,000	26,400	27,000	27,300
China(Mainland)	1,719,424	1,848,541	1,970,200	2,112,510	2,279,940	2,468,840	2,649,840	2,791,390	2,920,000	2,948,500	3,030,000
Taiwan	15,744	15,322	14,739	14,333	13,486	11,902	11,906	11,780	11,814	11,765	11,750
Japan	47,500	47,400	47,500	47,700	45,900	44,800	44,800	39,300	43,100	42,400	41,600
Iran	18,000	17,700	17,500	18,000	19,000	20,000	22,000	24,000	24,400	24,700	24,000
Malaysia	3300	3500	3350	3300	3200	3100	3100	3000	3000	3000	3000
Vietnam	131,487	133,000	129,000	127,000	124,027	127,000	131,000	134,000	134,000	134,000	134,000
Kenya	157,720	158,394	171,916	187,855	190,717	198,657	203,006	209,426	218,538	232,742	234,300
Uganda	23,800	24,530	26,000	28,700	29,400	32,538	33,700	36,000	37,500	43,940	45,000
Tanzania	22,722	22,721	22,721	22,721	22,721	22,700	22,700	22,721	22,722	22,609	22,600
Malawi	18,600	18,600	18,600	18,600	18,600	18,600	18,600	18,600	18,600	18,600	18,700
Zimbabwe	6000	5700	5800	5600	5400	5500	5500	5500	5500	5600	5640
Rest of Africa	40,880	35,023	39,698	31,095	36,390	41,170	41,320	48,530	47,800	51,920	53,410
Turkey	78,000	77,000	77,500	77,700	77,800	77,000	77,400	77,500	77,000	77,000	77,300
Argentina	38,000	38,900	40,210	40,610	40,680	40,730	40,730	40,750	41,000	41,000	41,400
Brazil	5300	5300	5340	5360	5000	4000	4000	3700	3600	3600	3500
Papua New Guinea	3830	3830	3830	3840	3850	3870	3880	3900	4000	4000	4000
Others	130,070	134,714	124,210	140,800	125,460	124,600	123,630	123,960	124,230	124,220	126,330
Grand total	33,419,420	3,556,169	3,680,621	3,832,876	3,988,292	4,196,305	4,394,930	4,550,928	4,702,488	4,794,320	4,917,527

Source: International Tea Committee except India (Tea Board).

Average yield per hectare										World (In kg.)
	2007	2008	2009–11	2010–12	2011–13	2012–14	2013–15	2014–16	2015–17	2016–18
India	1705	1750	1800	1883	2016	2085	2131	2153	2190	2221
Sri Lanka	1615	1692	1684	1756	1775	1791	1789	1658	1565	1487
Indonesia	1115	1038	1032	1174	1183	1106	1107	1089	1158	1152
Kenya	2477	2193	2106	2083	2044	2105	2089	2088	1986	2051
Bangladesh	1108	1087	1088	1118	1133	1189	1223	1243	1323	1388
Malawi	2587	2239	2710	2529	2437	2416	2361	2302	2297	2492
China (Mainland)	722	731	751	768	778	785	794	801	830	846

Source: International Tea Committee.

Area-wise production **North India**

	(Calendar year)										(In tons)
	2009	2010	2011	2012	2013	2014	2015	2016	2017	2018	2019
North India											
Darrang	–	–	–	–	–	–	–	–	–	–	–
Goalpara	–	–	–	–	–	–	–	–	–	–	–
Kamrup	–	–	–	–	–	–	–	–	–	–	–
Lakhimpur	–	–	–	–	–	–	–	–	–	–	–
Dibrugarh	–	–	–	–	–	–	–	–	–	–	–
Nowgong	–	–	–	–	–	–	–	–	–	–	–
Sibsagar	–	–	–	–	–	–	–	–	–	–	–
AssamValley †	445,126	428,736	538,820	537,250	573,980	561,560	585,820	618,340	627,980	645,140	670,780
Cachar	54,871	51,550	50,290	52,870	47,890	49,410	45,400	51,180	47,190	46,770	45,010
Assam	499,997	480,286	589,110	590,120	621,870	610,970	631,220	669,520	675,170	691,910	715,790
Darjeeling	10,742	8862	9140	8930	9130	8510	8760	8130	3210	7690	7920
Terai	74,041	76,355	114,040	113,660	125,850	131,790	133,960	143,700	161,720	160,200	176,730
Dooars	136,790	144,567	148,420	156,710	177,900	189,160	181,780	204,470	219,580	226,110	239,410
West Bengal	221,573	229,784	271,600	279,300	312,880	329,460	324,500	356,300	384,510	394,000	424,060
Tripura	–	–	–	–	–	–	–	–	–	–	–
Others!!	13,298	12,959	14,860	17,530	23,870	24,770	25,370	28,690	27,430	27,850	30,810
North India	734,868	723,029	875,570	886,950	958,620	965,200	981,090	1,054,510	1,087,110	1,113,760	1,170,660

– Denotes not available.
!! Including Bihar, Uttaranchal, Himachal Pradesh, Sikkim, Meghalaya, Mizoram, Manipur, Orissa, Arunachal Pradesh, and Nagaland.
†Includes: North Cachar and Karbi Anglong.
Source: Tea Board.

Area-wise production **South India**

	(Calendar year)										(In tons)
	2009	2010	2011	2012	2013	2014	2015	2016	2017	2018	2019
South India											
Kanyakumari	–	–	–	–	–	–	–	–	–	–	–
Tirunelveli	–	–	–	–	–	–	–	–	–	–	–
Madurai	–	–	–	–	–	–	–	–	–	–	–
Coimbatore	–	–	–	–	–	–	–	–	–	–	–
Nilgiris	–	–	–	–	–	–	–	–	–	–	–
Tamil Nadu	169,356	170,723	165,890	170,560	173,360	169,790	163,090	146,040	166,900	158,860	155,370
Palghat	–	–	–	–	–	–	–	–	–	–	–
Trichur	–	–	–	–	–	–	–	–	–	–	–
Trivandrum	–	–	–	–	–	–	–	–	–	–	–
Quilon	–	–	–	–	–	–	–	–	–	–	–
Kottayam	–	–	–	–	–	–	–	–	–	–	–
Idukki	–	–	–	–	–	–	–	–	–	–	–
Wynaad	–	–	–	–	–	–	–	–	–	–	–
Kerala	68,964	66,754	68,830	63,100	62,840	65,580	57,970	61,400	62,350	60,820	59,050
Chikmagalur	–	–	–	–	–	–	–	–	–	–	–
Coorg	–	–	–	–	–	–	–	–	–	–	–
Hassan	–	–	–	–	–	–	–	–	–	–	–
Karnataka	5811	5897	5430	5720	5590	6740	6510	5410	5400	5190	4620
South India	244,131	243,374	240,150	239,830	241,790	241,110	227,570	212,850	234,650	224,870	219,040
All India	978,999	966,403	1,115,720	1,126,330	1,200,410	1,207,310	1,208,660	1,267,360	1,321,760	1,338,630	1,389,700

– Denotes not available.
Source : Tea Board.

Area-wise monthly production of Indian tea.

(In tons)

Months		Assam* valley	Cachar	Darjeeling	Dooars	Terai	Others	Total North India	Tamil nadu	Kerala	Karnataka	Total South India	Total India
January	2019	–	–	–	–	–	–	–	9510	4450	260	14,220	14,220
	2018	850	570	–	1030	2670	170	5290	8130	3990	270	12,390	17,680
February	2019	350	90	10	2110	1920	160	4640	7170	2750	240	10,160	14,800
	2018	400	140	10	850	430	20	1850	7940	3420	270	11,630	13,480
March	2019	31,430	1480	220	14,200	10,680	1480	59,490	10,400	4350	350	15,100	74,590
	2018	20,410	1540	210	13,460	10,260	840	46,720	9720	4200	400	14,320	61,040
April	2019	41,910	3010	1300	13,610	8390	2150	70,370	10,450	3280	320	14,050	84,420
	2018	45,040	3280	1070	9690	4760	1620	65,460	15,320	6670	530	22,520	87,980
May	2019	61,850	4670	780	24,060	18,510	2930	112,800	15,110	6470	410	21,990	134,790
	2018	52,570	4020	820	20,790	15,470	2710	96,380	18,570	6970	600	26,140	122,520
June	2019	70,330	4920	950	25,520	20,290	3850	125,860	17,990	6530	480	25,000	150,860
	2018	73,600	4950	770	25,060	18,630	2960	125,970	17,020	5540	600	23,160	149,130
July	2019	91,310	5710	1060	31,180	21,380	3700	154,340	15,400	5830	500	21,730	176,070
	2018	87,580	6130	1120	28,470	19,050	3550	145,900	12,960	3400	250	16,610	162,510
August	2019	91,470	5900	1440	31,760	22,120	4430	157,120	10,250	4210	310	14,770	171,890
	2018	108,450	7040	1350	33,620	22,760	4550	177,770	10,280	1850	170	12,300	190,070
September	2019	102,380	6100	1150	31,200	22,740	4240	167,810	13,040	3590	280	16,910	184,720
	2018	94,510	6060	1000	28,730	20,990	4070	155,360	13,660	5390	380	19,430	174,790
October	2019	92,740	6060	610	27,780	20,790	3700	151,680	17,750	6890	590	25,230	176,910
	2018	91,520	6580	820	31,850	21,140	3850	155,760	17,870	8450	730	27,050	182,810
November	2019	65,630	4800	340	25,380	18,790	2700	117,640	15,690	5680	380	21,750	139,390
	2018	56,240	4670	410	22,340	15,390	2460	101,510	14,030	5070	490	19,590	121,100
December	2019	21,380	2270	60	12,610	11,120	1470	48,910	12,610	5020	500	18,130	67,040
	2018	13,970	1790	110	10,220	8650	1050	35,790	13,360	5870	500	19,730	55,520
Upto December	2019	670,780	45,010	7920	239,410	176,730	30,810	1,170,660	155,370	59,050	4620	219,040	1,389,700
	2018	645,140	46,770	7690	226,110	160,200	27,850	1,113,760	158,860	60,820	5190	224,870	1,338,630

N. B.: Total North India figures include production of Tripura, Bihar, Uttaranchal, Himachal Pradesh, Manipur, Arunachal Pradesh, Nagaland, Mizoram, Meghalaya, Orissa and Sikkim.

*Includes: KarbiAnglong, – Denotes Nil.

Source: Tea Board.

Production of tea in India by different method of manufacture.

(Quantity in M. Kgs.)

	North India				South India			All India			
Year	CTC	Orthodox	Darjeeling	Total	CTC	Orthodox	Total	CTC	Orthodox	Darjeeling	Total
1997	549.5	37.1	10.1	604.7	150.6	54.2	205.3	700.1	91.3	10.1	810.0
1998	594.1	59.5	10.3	670.7	144.1	58.8	203.4	738.2	118.3	10.3	874.1
1999	576.0	32.5	8.7	623.2	172.1	30.2	202.7	748.1	62.7	8.7	825.9
2000	590.1	35.7	9.3	640.7	170.3	35.3	206.2	760.4	71.0	9.3	846.9
2001	592.1	44.4	9.8	650.8	167.4	34.8	203.1	759.5	79.2	9.8	853.9
2002	570.4	48.4	9.2	631.8	174.7	31.1	206.7	745.1	79.5	9.2	838.5
2003	607.9	26.9	9.6	648.3	191.6	37.2	229.8	799.5	64.1	9.6	878.1
2004	617.8	28.9	10.1	662.2	196.8	32.5	230.8	814.6	61.4	10.1	893.0
2005	667.0	32.6	11.3	718.4	182.4	43.3	227.6	849.4	75.9	11.3	946.0
2006	702.1	30.8	10.9	753.2	191.3	35.5	228.6	893.4	66.3	10.9	981.8
2007	701.4	44.5	10.0	764.7	186.5	34.1	221.7	887.9	78.6	10.0	986.4
2008	676.0	50.0	11.0	737.0	207.0	38.0	245.0	883.0	88.0	11.0	982.0
2009	674.0	53.0	10.0	737.0	202.0	40.0	242.0	876.0	93.0	10.0	979.0
2010	659.0	55.0	9.0	723.0	200.0	42.0	242.0	859.0	97.0	9.0	965.0
2011	797.0	69.0	9.0	875.0	193.0	44.0	237.0	990.0	113.0	9.0	1112.0
2012	823.0	56.0	9.0	888.0	193.0	43.0	236.0	1016.0	99.0	9.0	1124.0
2013	888.0	59.0	9.0	956.0	191.0	46.0	237.0	1079.0	105.0	9.0	1193.0
2014	915.0	40.0	9.0	964.0	193.0	44.0	237.0	1108.0	84.0	9.0	1201.0
2015	915.7	56.7	8.8	981.1	183.0	44.6	227.6	1098.7	101.2	8.8	1208.7
2016	981.0	65.4	8.1	1054.5	167.5	45.4	212.9	1148.5	110.7	8.1	1267.4
2017	1007.6	76.3	3.2	1087.1	182.7	52.0	234.7	1190.2	128.3	3.2	1321.8
2018	1033.2	72.9	7.7	1113.8	175.6	49.3	224.9	1208.8	122.1	7.7	1338.6

Source: International Tea Committee except 2015 to 2018 "Tea Board".

Monthly production Figures **North India**

(In tons)

Months	2009	2010	2011	2012	2013	2014	2015	2016	2017	2018	2019
January	10,123	8700	3960	2690	2850	4960	4180	3370	6050	5290	—
February	3081	1527	950	920	920	1790	2250	2290	2030	1850	4640
YTD	13,204	10,227	4910	3610	3770	6750	6430	5660	8080	7140	4640
March	27,341	29,829	35,070	26,390	32,660	36,130	27,070	55,310	41,490	46,720	59,490
YTD	40,545	40,056	39,980	30,000	36,430	42,880	33,500	60,970	49,570	53,860	64,130
April	39,684	48,043	52,420	49,090	54,910	42,080	53,930	51,630	65,860	65,460	70,370
YTD	80,229	88,099	92,400	79,090	91,340	84,960	87,430	112,600	115,430	119,320	134,500
May	50,322	47,406	78,870	78,850	79,030	69,290	95,590	87,500	98,330	96,380	112,800
YTD	130,551	135,505	171,270	157,940	170,370	158,250	183,020	200,100	213,760	215,700	247,300
June	91,875	76,796	103,460	104,560	112,180	113,430	100,740	122,140	121,390	125,970	125,860
YTD	222,426	212,301	274,730	262,500	282,550	267,680	283,760	322,240	335,150	341,670	373,160
July	106,429	102,811	119,350	112,530	138,250	140,970	138,640	131,230	142,810	145,900	154,340
YTD	328,855	315,112	394,080	375,030	420,800	408,650	422,400	453,470	477,960	487,570	527,500
August	114,365	107,825	124,330	133,620	140,830	149,840	139,970	140,510	156,080	177,770	157,120
YTD	443,220	422,937	518,410	508,650	561,630	558,490	562,370	593,980	634,040	665,340	684,620
September	78,796	89,708	129,620	128,760	132,190	137,540	136,960	169,680	133,750	155,360	167,810
YTD	522,016	512,645	648,030	637,410	693,820	696,030	699,330	763,660	767,790	820,700	852,430
October	112,178	99,242	119,210	119,110	140,090	144,640	144,890	132,890	162,060	155,760	151,680
YTD	634,194	611,887	767,240	756,520	833,910	840,670	844,220	896,550	929,850	976,460	1,004,110
November	64,157	70,823	74,860	91,250	85,440	82,760	94,180	109,740	102,290	101,510	117,640
YTD	698,351	682,710	842,100	847,770	919,350	923,430	938,400	1,006,290	1,032,140	1,077,970	1,121,750
December	36,517	40,319	33,470	39,180	39,270	41,770	42,690	48,220	54,970	35,790	48,910
YTD	734,868	723,029	875,570	886,950	958,620	965,200	981,090	1,054,510	1,087,110	1,113,760	1,170,660

Source: Tea Board.

Monthly production Figures — **South India**

(In tons)

Months	2009	2010	2011	2012	2013	2014	2015	2016	2017	2018	2019
January	11,447	18,390	18,760	16,200	19,120	17,100	17,160	14,580	13,110	12,390	14,220
February	12,194	16,443	15,630	13,220	12,320	13,930	13,950	14,840	11,560	11,630	10,160
YTD	23,641	34,833	34,390	29,420	31,440	31,030	31,110	29,420	24,670	24,020	24,380
March	17,695	19,175	20,210	14,900	15,190	17,520	16,690	15,390	14,670	14,320	15,100
YTD	41,336	54,008	54,600	44,320	46,630	48,550	47,800	44,810	39,340	38,340	39,480
April	22,906	20,198	21,190	21,900	20,260	16,500	24,830	17,010	26,550	22,520	14,050
YTD	64,242	74,206	75,790	66,220	66,890	65,050	72,630	61,820	65,890	60,860	53,530
May	21,052	25,221	26,960	24,570	24,640	25,770	26,100	14,970	25,180	26,140	21,990
YTD	85,294	99,427	102,750	90,790	91,530	90,820	98,730	76,790	91,070	87,000	75,520
June	26,175	27,241	23,490	26,640	24,680	30,210	24,770	25,010	27,010	23,160	25,000
YTD	111,469	126,668	126,240	117,430	116,210	121,030	123,500	101,800	118,080	110,160	100,520
July	20,558	20,383	19,000	20,490	16,880	23,940	18,560	19,300	19,470	16,610	21,730
YTD	132,027	147,051	145,240	137,920	133,090	144,970	142,060	121,100	137,550	126,770	122,250
August	20,011	14,909	14,910	17,280	14,640	14,160	17,270	18,190	20,160	12,300	14,770
YTD	152,038	161,960	160,150	155,200	147,730	159,130	159,330	139,290	157,710	139,070	137,020
September	22,672	18,392	16,910	15,350	24,170	19,790	18,500	20,320	18,460	19,430	16,910
15,300YTD	174,710	180,352	177,060	170,550	171,900	178,920	177,830	159,610	176,170	158,500	153,930
October	21,459	21,495	23,330	25,380	24,980	25,430	18,540	18,260	21,170	27,050	25,230
YTD	196,169	201,847	200,390	195,930	196,880	204,350	196,370	177,870	197,340	185,550	179,160
November	26,374	23,100	22,090	22,340	27,040	20,150	16,600	19,200	20,890	19,590	21,750
YTD	222,543	224,947	222,480	218,270	223,920	224,500	212,970	197,070	218,230	205,140	200,910
December	21,588	18,427	17,670	21,110	17,870	17,610	14,600	15,780	16,420	19,730	18,130
YTD	244,131	243,374	240,150	239,380	241,790	242,110	227,570	212,850	234,650	224,870	219,040

Source: Tea Board.

Monthly production Figures — **All India**

(In tons)

Months	2009	2010	2011	2012	2013	2014	2015	2016	2017	2018	2019
January	21,570	27,090	22,720	18,890	21,970	22,060	21,340	17,950	19,160	17,680	14,220
February	15,275	17,970	16,580	14,140	13,240	15,720	16,200	17,130	13,590	13,480	14,800
YTD	36,845	45,060	39,300	33,030	35,210	37,780	37,540	35,080	32,750	31,160	29,020
March	45,036	49,004	55,280	41,290	47,850	53,650	43,760	70,700	56,160	61,040	74,590
YTD	81,881	94,064	94,580	74,320	83,060	91,430	81,300	105,780	88,910	92,200	103,610
April	62,590	68,241	73,610	70,990	75,170	58,580	78,760	68,640	92,410	87,980	84,420
YTD	144,471	162,305	168,190	145,310	158,230	150,010	160,060	174,420	181,320	180,180	188,030
May	71,374	72,627	105,830	103,420	103,670	95,060	121,690	102,470	123,510	122,520	134,790
YTD	215,845	234,932	274,020	248,730	261,900	245,070	281,750	276,890	304,830	302,700	322,820
June	118,050	104,037	126,950	131,200	136,860	143,640	125,510	147,150	148,400	149,130	150,860
YTD	333,895	338,969	400,970	379,930	398,760	388,710	407,260	424,040	453,230	451,830	473,680
July	126,987	123,194	138,350	133,020	155,130	164,910	157,200	150,530	162,280	162,510	176,070
YTD	460,882	462,163	539,320	512,950	553,890	553,620	564,460	574,570	615,510	614,340	649,750
August	134,376	122,734	139,240	150,900	155,470	164,000	157,240	158,700	176,240	190,070	171,890
YTD	595,258	584,897	678,560	663,850	709,360	717,620	721,700	733,270	791,750	804,410	821,640
September	101,468	108,100	146,530	144,110	156,360	157,330	155,460	190,000	152,210	174,790	184,720
YTD	696,726	692,997	825,090	807,960	865,720	874,950	877,160	923,270	943,960	979,200	1,006,360
October	133,637	120,737	142,540	144,490	165,070	170,070	163,430	151,150	183,230	182,810	176,910
YTD	830,363	813,734	967,630	952,450	1,030,790	1,045,020	1,040,590	1,074,420	1,127,190	1,162,010	1,183,270
November	90,531	93,923	96,950	113,590	112,480	102,910	110,780	128,940	123,180	121,100	139,390
YTD	920,894	907,657	1,064,580	1,066,040	1,143,270	1,147,930	1,151,370	1,203,360	1,250,370	1,283,110	1,322,660
December	58,105	58,746	51,140	60,290	57,140	59,380	57,290	64,000	71,390	55,520	67,040
YTD	978,999	966,403	1,115,720	1,126,330	1,200,410	1,207,310	1,208,660	1,267,360	1,321,760	1,338,630	1,389,700

Source: Tea Board.

Monthly production: Big growers and small growers **All India**

(In tons)

	North India				South India				All India			
	Big growers		Small growers		Big growers		Small growers		Big growers		Small growers	
Months	**2018**	**2019**	**2018**	**2019**	**2018**	**2019**	**2018**	**2019**	**2018**	**2019**	**2018**	**2019**
January	1730	—	3560	—	7240	7950	5150	6270	8970	7950	8710	6270
February	1020	1780	830	2860	6460	5510	5170	4650	7480	7290	6000	7510
YTD	2750	1780	4390	2860	13,700	13,460	10,320	10,920	16,450	15,240	14,710	13,780
March	21,620	27,060	25,100	32,430	8220	8220	6100	6880	29,840	35,280	31,200	39,310
YTD	24,370	28,840	29,490	35,290	21,920	21,680	16,420	17,800	46,290	50,520	45,910	53,090
April	35,530	37,900	29,930	32,470	12,350	6670	10,170	7380	47,880	44,570	40,100	39,850
YTD	59,900	66,740	59,420	67,760	34,270	28,350	26,590	25,180	94,170	95,090	86,010	92,940
May	46,260	53,320	50,120	59,480	13,250	12,310	12,890	9680	59,510	65,630	63,010	69,160
YTD	106,160	120,060	109,540	127,240	47,520	40,660	39,480	34,860	153,680	160,720	149,020	162,100
June	65,040	62,220	60,930	63,640	11,840	12,450	11,320	12,550	76,880	74,670	72,250	76,190
YTD	171,200	182,280	170,470	190,880	59,360	53,110	50,800	47,410	230,560	235,390	221,270	238,290
July	77,050	80,130	68,850	74,210	7020	10,650	9590	11,080	84,070	90,780	78,440	85,290
YTD	248,250	262,410	239,320	265,090	66,380	63,760	60,390	58,490	314,630	326,170	299,710	323,580
August	97,810	81,900	79,960	75,220	4810	7870	7490	6900	102,620	89,770	87,450	82,120
YTD	346,060	344,310	319,280	340,310	71,190	71,630	67,880	65,390	417,250	415,940	387,160	405,700
September	82,580	88,710	72,780	79,100	10,040	7600	9390	9310	92,620	96,310	82,170	88,410
YTD	428,640	433,020	392,060	419,410	81,230	79,230	77,270	74,700	509,870	512,250	469,330	494,110
October	83,490	80,550	72,270	71,130	15,670	12,690	11,380	12,540	99,160	93,240	83,650	83,670
YTD	512,130	513,570	464,330	490,540	96,900	91,620	88,650	87,240	609,030	605,490	552,980	577,780
November	48,770	57,660	52,740	59,980	9390	10,960	10,200	10,790	58,160	68,620	62,940	70,770
YTD	560,900	571,230	517,070	550,520	106,290	102,880	98,850	98,030	667,190	674,110	615,920	648,550
December	14,560	21,280	21,230	27,630	10,510	9230	9220	8900	25,070	30,510	30,450	36,530
YTD	575,460	592,510	538,300	578,150	116,800	112,110	108,070	106,930	692,260	704,620	646,370	685,080

Source: Tea Board.

Monthly production figures **Sri Lanka**

(In tons)

Months	2009	2010	2011	2012	2013	2014	2015	2016	2017	2018	2019
January	17,653	26,898	20,698	22,409	23,207	25,425	23,261	25,082	21,256	24,271	23,208
February	12,558	23,508	21,657	21,985	24,878	22,475	25,456	22,900	18,462	21,328	21,719
YTD	30,211	50,406	42,355	44,394	48,085	47,949	48,717	47,982	39,718	45,599	44,927
March	18,163	20,498	33,244	27,208	32,197	25,569	30,271	22,067	26,434	28,628	28,263
YTD	48,374	70,904	75,599	71,602	80,282	73,552	79,089	70,049	66,152	74,227	73,190
April	29,173	28,415	28,496	29,330	33,596	28,958	31,776	26,471	32,790	28,018	23,863
YTD	77,547	99,319	104,095	100,932	113,878	102,510	110,865	96,520	98,942	102,245	97,053
May	25,543	30,291	33,647	31,481	33,669	39,045	32,240	30,627	29,459	32,928	34,114
YTD	103,090	129,610	137,742	135,629	148,270	141,555	143,105	127,147	128,401	135,173	131,167
June	23,623	29,509	28,796	27,636	24,699	30,710	28,918	25,360	27,103	23,032	27,302
YTD	126,713	159,119	166,538	163,265	173,771	172,551	172,616	152,507	155,504	158,205	158,469
July	25,118	26,407	24,195	24,893	22,668	28,794	25,790	21,326	25,597	24,048	25,596
YTD	151,831	185,526	190,733	188,158	196,439	201,306	198,406	173,833	181,101	182,253	184,065
August	23,875	25,154	25,362	26,263	26,584	24,772	26,042	23,011	24,684	21,436	26,924
YTD	175,706	210,680	216,095	214,421	223,023	226,078	225,589	196,844	205,785	203,689	210,989
September	25,512	25,046	22,925	28,181	25,027	28,855	26,885	19,809	25,460	19,575	21,737
YTD	201,218	235,726	239,020	242,602	248,495	255,747	252,698	216,653	231,245	223,264	232,726
October	26,730	26,642	24,844	26,998	30,250	28,011	26,735	18,566	25,469	29,709	21,002
YTD	227,948	262,368	263,864	269,600	278,745	283,758	279,433	235,219	256,714	252,973	253,728
November	28,291	26,668	31,822	30,909	29,766	29,700	25,204	27,313	24,773	24,804	24,450
YTD	256,239	289,036	295,686	300,509	308,511	313,458	304,637	262,532	281,487	277,777	278,178
December	25,196	25,690	25,710	26,651	30,212	24,307	23,518	27,085	23,503	26,229	21,946
YTD	281,435	314,726	321,396	328,397	340,229	338,032	328,960	292,574	307,720	304,006	300,134

Source: International Tea Committee.

Monthly production figures **Kenya**

(In tons)

Months	2009	2010	2011	2012	2013	2014	2015	2016	2017	2018	2019
January	25,484	37,713	35,999	36,205	45,390	44,970	41,600	50,300	32,991	40,834	48,386
February	21,539	34,834	26,711	18,412	38,500	33,770	24,280	43,970	22,605	27,939	31,445
YTD	47,023	72,547	62,710	54,617	83,890	78,740	65,880	94,270	55,596	68,773	79,831
March	18,779	39,175	22,459	17,859	33,368	33,340	15,688	45,330	34,498	30,986	26,462
YTD	65,802	111,722	85,169	72,476	117,258	112,080	81,568	139,600	90,094	99,759	106,293
April	18,343	35,857	31,482	18,120	38,230	39,970	23,840	37,570	31,458	44,580	26,131
YTD	84,145	147,579	116,651	90,596	155,488	152,050	105,408	177,170	121,552	144,339	132,424
May	29,803	35,618	32,856	37,383	39,600	41,186	37,523	36,573	38,822	43,356	37,750
YTD	113,948	183,197	149,507	127,979	195,088	193,242	142,977	213,743	160,374	187,695	170,174
June	25,336	29,815	28,955	30,200	30,530	31,945	32,286	35,603	40,538	43,299	42,434
YTD	139,284	213,012	178,462	158,179	225,618	225,187	175,263	249,346	200,912	230,994	212,608
July	21,510	24,401	26,343	24,306	26,229	30,790	30,940	29,285	31,565	35,278	31,458
YTD	160,794	237,413	204,805	182,485	251,847	255,977	206,203	278,631	232,477	266,272	244,066
August	21,231	23,177	24,471	32,000	26,338	26,756	28,410	29,462	32,693	37,433	37,201
YTD	182,025	260,590	229,276	214,485	278,185	282,733	234,613	308,093	265,170	303,705	281,267
September	27,434	28,883	30,493	33,549	32,055	33,321	36,484	36,787	38,386	42,531	35,533
YTD	209,459	289,473	259,769	248,034	310,240	316,054	271,097	344,880	303,556	346,236	316,800
October	32,760	34,140	39,926	40,235	44,283	45,368	41,340	41,340	43,420	49,283	46,300
YTD	242,219	323,613	299,695	288,269	354,523	361,422	312,437	386,220	346,976	395,519	363,100
November	35,866	37,063	36,825	39,977	36,211	38,610	40,382	39,903	45,374	45,649	45,087
YTD	278,085	360,676	336,520	328,164	390,734	400,032	352,819	426,123	392,350	441,168	408,187
December	36,114	38,330	41,393	41,401	41,719	45,071	46,387	45,100	47,507	51,830	50,666
YTD	314,199	399,006	377,913	369,562	432,453	445,103	399,211	473,011	439,858	492,998	458,853

Source: International Tea Committee.

Monthly production figures **Bangladesh**

(In tons)

Months	2009	2010	2011	2012	2013	2014	2015	2016	2017	2018	2019
January	338	238	397	181	189	250	160	51	157	331	495
February	24	31	19	185	46	56	13	6	16	38	114
YTD	362	269	416	366	235	306	173	57	173	369	609
March	510	27	676	532	500	501	357	2001	1152	1556	1917
YTD	872	296	1092	898	805	807	530	2058	1325	1925	2526
April	2679	1906	3506	3313	2083	2703	2854	4733	4382	3505	6110
YTD	3551	2202	4598	4044	2888	3510	3384	6791	5707	5430	8636
May	4000	5030	4482	5602	5348	4645	5351	5462	5934	6334	7645
YTD	7551	7232	9080	9646	8236	8155	8735	12,253	11,641	11,764	16,281
June	6655	6335	6586	7889	6948	8100	5764	10,415	8584	7634	11,667
YTD	14,206	13,567	15,666	17,535	15,184	16,255	14,499	22,668	20,225	19,398	27,948
July	7,9 93	8043	8282	8123	10,112	8930	9378	9891	11,206	10,985	11,104
YTD	22,199	21,610	23,948	25,658	25,296	25,185	23,877	32,559	31,431	30,383	39,052
August	9086	8524	8194	9210	9494	10,300	10,360	10,554	10,648	12,477	13,505
YTD	31,285	30,134	32,142	34,868	34,790	35,477	34,237	43,113	42,079	42,860	52,557
September	7695	7657	8737	8765	9287	9005	10,040	11,914	9784	12,008	13,382
YTD	38,980	37,791	40,879	43,633	44,077	44,464	44,277	55,027	51,863	54,868	65,939
October	10,012	8850	8287	8608	9673	10,363	9500	10,412	11,464	11,863	13,407
YTD	48,992	46,641	49,166	52,241	53,750	54,827	53,777	65,439	63,327	66,731	79,346
November	4998	6964	6475	6542	7174	5816	7750	8843	9075	9147	10,300
YTD	53,990	53,605	55,641	58,783	60,924	60,643	61,527	74,282	72,402	75,878	89,646
December	5252	5317	3797	3372	5336	3836	4820	5083	6547	6256	6423
YTD	59,242	58,922	59,438	62,155	66,260	64,479	66,347	79,365	78,949	82,134	96,069

Source: International Tea Committee.

Monthly production figures **Malawi**

(In tons)

Months	2009	2010	2011	2012	2013	2014	2015	2016	2017	2018	2019
January	7102	6275	6814	6384	6048	8200	6483	5599	5224	6339	7446
February	6874	7350	6792	7337	5120	6144	5082	5773	7008	6759	7283
YTD	13,976	13,625	13,606	13,721	11,168	14,344	11,565	11,372	12,232	13,098	14,729
March	6940	7995	7259	7595	6555	7443	6923	7740	7132	7685	6442
YTD	20,916	21,620	20,865	21,316	17,723	21,787	18,488	19,112	19,364	20,783	21,171
April	5791	6877	6254	5201	5544	6371	6354	5896	5705	6522	7375
YTD	26,707	28,497	27,119	26,517	23,267	28,158	24,842	25,008	25,069	27,305	28,546
May	4139	4353	4022	3825	4661	4575	4121	3394	4298	4938	4490
YTD	30,846	32,850	31,141	30,342	27,928	32,733	28,963	28,402	29,367	32,243	33,036
June	3275	2663	1029	1575	2348	2195	1674	1502	2176	1901	1618
YTD	34,121	35,513	32,170	31,917	30,276	34,928	30,637	29,904	31,543	34,144	34,654
July	1800	1713	1029	1149	1654	1382	1124	886	1475	1250	1359
YTD	35,921	37,226	33,199	33,066	31,930	36,310	31,761	30,790	33,018	35,394	36,013
August	1900	1466	1024	1207	1800	1630	830	1779	1498	1666	1667
YTD	37,821	38,692	34,223	34,273	33,730	37,940	32,591	32,569	34,516	37,060	37,680
September	2988	3448	1646	1403	2663	2715	1309	2805	1838	3572	2110
YTD	40,809	42,140	35,869	35,676	36,393	40,655	33,900	35,374	36,354	40,632	39,790
October	2464	2890	2198	1312	3245	2344	980	1435	1472	2199	2302
YTD	43,273	45,030	38,067	36,988	39,638	42,999	34,880	36,809	37,826	42,831	42,092
November	1652	878	3312	971	2757	1072	750	741	2106	1409	1117
YTD	44,925	45,908	41,379	37,959	42,395	44,071	35,630	37,550	39,932	44,240	43,209
December	7633	5682	4407	4532	4065	1784	3813	5577	5650	6346	4983
YTD	52,558	51,590	47,058	42,490	46,463	45,855	39,443	43,127	45,582	50,586	48,192

Source: International Tea Committee.

Production										World
	(Calendar year)									(In tons)
	2010	2011	2012	2013	2014	2015	2016	2017	2018	2019
India	966,403	1,115,720	1,126,330	1,200,410	1,207,310	1,208,660	1,267,360	1,321,760	1,338,630	1,389,700
Sri Lanka	331,427	328,632	328,397	340,026	338,032	328,964	292,574	307,720	304,006	300,134
Indonesia	151,012	142,342	137,769	136,856	135,721	129,293	126,524	134,000	131,000	128,800
Bangladesh	61,620	65,908	62,155	66,260	64,480	66,347	79,364	78,949	82,134	96,069
China (Mainland)	1,475,060	1,623,214	1,789,753	1,924,457	2,095,717	2,248,999	2,404,947	2,496,412	2,610,393	2,700,000
Taiwan	17,467	17,309	14,902	14,717	15,200	14,405	13,018	13,443	13,300	13,900
Japan	83,000	82,100	85,900	82,800	81,330	76,400	77,100	78,800	81,500	81,700
Iran	16,800	16,000	14,700	14,000	13,600	13,000	20,700	21,400	21,000	20,000
Nepal	17,438	18,310	20,588	21,076	23,187	23,000	23,000	24,200	24,800	25,000
Vietnam	175,000	178,000	174,028	180,325	175,000	170,000	180,000	175,000	163,000	158,000
Turkey	231,149	246,120	230,559	235,209	246,458	258,541	253,312	255,404	252,000	253,000
CIS	7850	7980	8190	8280	8390	8490	8800	9020	9630	9730
Kenya	399,006	377,912	369,562	432,453	445,105	399,211	473,011	439,858	492,999	458,853
Uganda	59,140	54,178	57,939	60,970	65,373	58,588	55,736	53,887	71,567	57,000
Tanzania	31,646	32,775	32,282	32,123	36,115	31,658	29,112	31,814	35,170	33,659
Malawi	51,591	47,056	42,490	46,463	45,855	39,447	43,127	45,582	50,588	48,192
Zimbabwe	14,293	14,622	13,000	13,300	14,000	15,200	15,500	15,700	15,200	14,900
Rest of Africa	53,789	56,554	59,100	59,930	60,351	63,819	62,850	67,017	68,500	72,346
Argentina	92,417	92,892	82,813	80,423	82,313	83,000	84,000	82,000	80,000	77,000
Brazil	5400	5500	5700	5100	4500	4000	4300	3600	3400	3400
Papua New Guinea	6800	6400	6200	6300	6400	6600	6800	6900	7000	6700
Others	32,370	32,860	33,180	33,590	34,200	33,900	34,700	35,520	56,955	57,700
Grand total	4,162,527	4,562,384	4,695,537	4,995,068	5,198,637	5,281,522	5,555,835	5,697,986	5,912,772	6,005,783

Source: International Tea Committee except India (Tea Board).

Production of green tea **World**

	(Calendar year)									(In tons)
	2009	2010	2011	2012	2013	2014	2015	2016	2017	2018
India	14,060	16,000	12,350	11,080	17,240	14,850	19,080	18,630	20,070	19,550
Bangladesh	250	260	240	250	270	260	260	270	270	280
Sri Lanka	2306	3285	2988	3032	3725	3195	2879	2374	2662	2601
Indonesia	32,200	33,000	31,000	33,000	31,600	31,400	28,000	28,000	27,600	27,000
China (Mainland)	1,006,302	1,046,382	1,137,646	1,247,827	1,313,362	1,416,238	1,494,644	1,588,005	1,678,971	1,722,385
Japan	85,957	82,988	82,076	85,859	82,755	81,277	76,200	76,900	78,600	81,300
Korea, rep.	4100	4140	4100	4120	4300	4400	4500	4600	4640	4800
Taiwan	16,500	17,100	16,700	14,200	14,000	14,400	13,700	12,700	13,100	13,000
Vietnam	72,000	63,000	80,100	83,300	88,700	95,500	93,500	90,000	97,000	90,000
Georgia	580	590	600	580	570	580	570	600	600	600
Russian Federation	630	640	620	630	620	600	600	620	600	620
Argentina	1650	1800	1900	1870	2200	2400	2200	1800	1700	1700
Brazil	580	600	600	620	640	600	570	600	530	470
Grand total	1,237,115	1,269,785	1,370,920	1,487,368	1,559,982	1,665,700	1,736,703	1,788,094	1,926,343	1,964,306

Source: International Tea Committee.

1.4 Excerpts of J. Thomas statistics, Kolkatta 2019 report, and UPASI Coonoor, planters Chronicle, August 2020

The crop production and marketing are governed by many factors such as areas increase/decrease, climate, seasonal—vagaries of monsoon, storms, floods, demand and supply, procedural formalities in transportation storage, marketing procedures, and formalities contribute to the fluctuations in supply and demand, and consequently the price variation, thereby the value realization. Sustainability is an important phenomenon in crop production, quality, and value realization. The package of practices, incidence of pests and diseases, field handling, processing, quality, marketing, value addition, and thereby value these realization and play a role on the prosperity/sustainability of tea plantations globally.

A lot of fine tuning, upgradation, reforms in fields, handling, processing, value addition, and marketing are required to be attended to carry forward the tea industry for the success and happiness/glory of the industry, producer, consumers, and marketers. The population working, managing, and coordinating the steps from production to marketing/consumers play a vital role in the sustainability. Timely restoration measures with innovations, improved packaging practices, innovative and dependable bioinputs at reasonable prices with management guidelines, and financial flow, accountability, dedication, and management for implementing the improved health and welfare measures of the work force and dependent populations will pave the way for sustainable industries, with a congenial, egoistic emerging society, compassionate ecofriendly management, and an understanding for the prosperous well-being of tea plantations, globally.

The following action and developments within the two to three years' timeframe will restore vertical growth, doubling the productivity with quality, ensuring a prosperous and healthy industry; a win—win situation for all the stakeholders of the tea industry globally (Figures 1.1 and 1.2).

1. Restore ecofriendly soil, plant, and environmental biospheres, along with biodynamics with judicious inputs regulating gaps in natural farming, diversification, value addition, and aggressive marketing, satisfying all the stakeholders and providing a cleaner, tastier, and healthier beverage for happiness and satisfaction.
2. The existing potent cultivars are improved further with breeding, rationalizing the bioinputs, advanced management practices, inputs, and

judicious, need-based spraying for sustainable productivity with quality and value realization.

3. Provide and maintain the plantations's ecofriendly, congenial environment, economic, productivity, and refined packages of practices sustained with cash flow, welfare for the society of the depending population, and all stakeholders coordinated by the science, administration, finance, environment, soil—water resources, sustainability, quality, prosperity, and happy living.

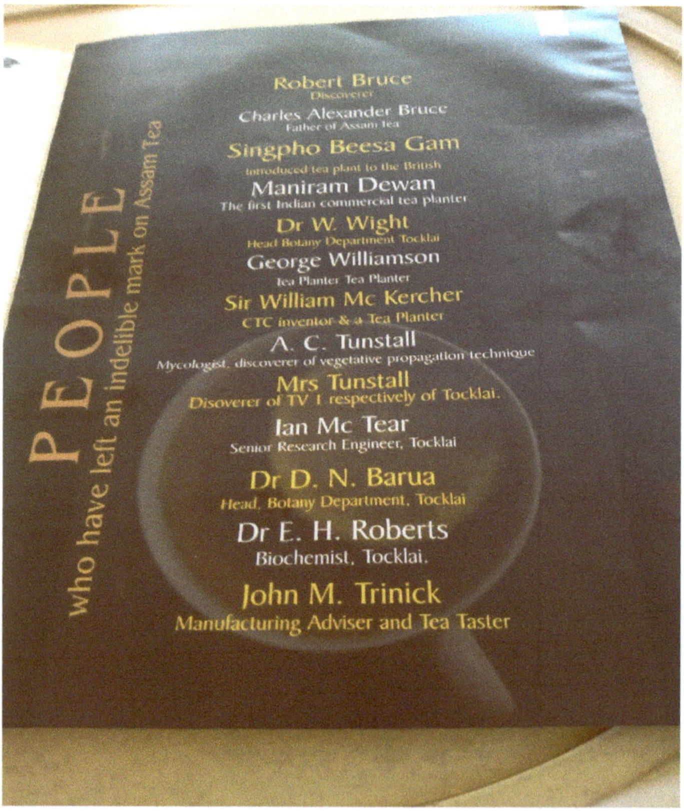

Figure 1.1 History of Tea: Early contributors of AssamTea. *Source: ATPA Year Book, 2017–18, 81stAGM, souvenir.*

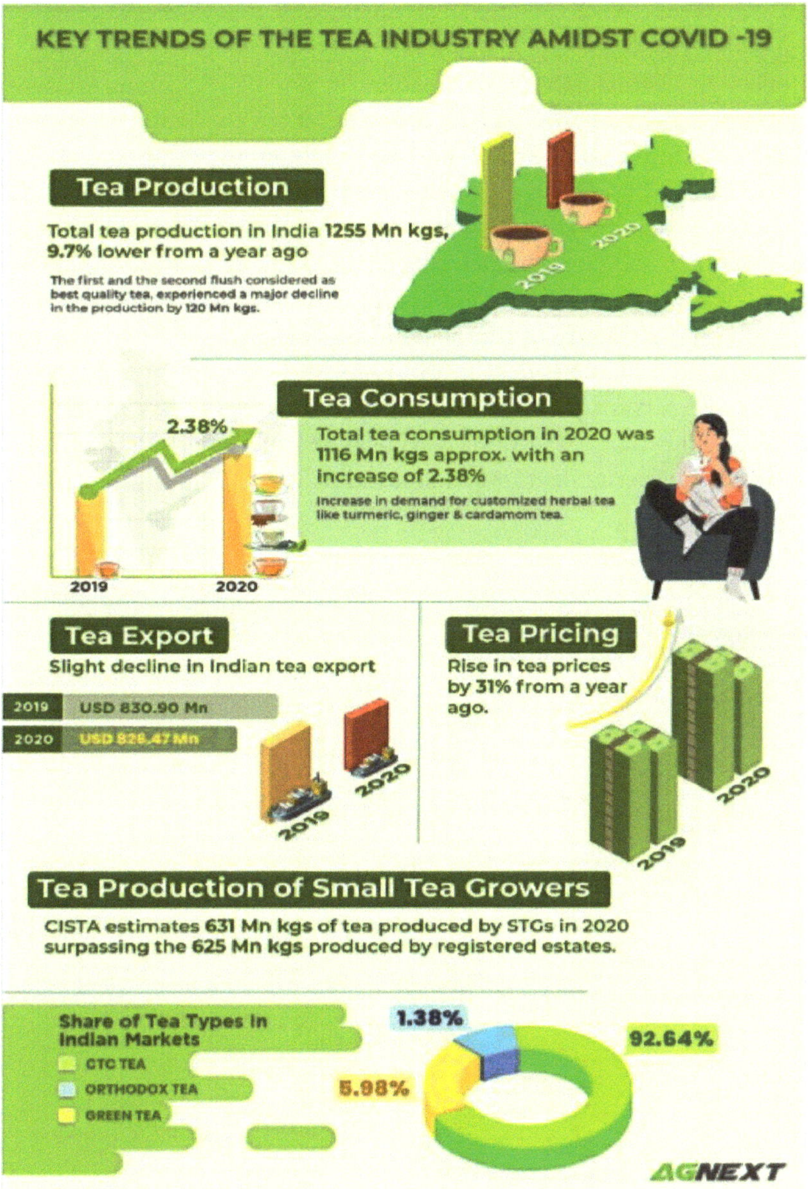

Figure 1.2 Key trends of the tea Industry amidst COVID-19.

1.5 Impact of pandemic on Indian tea

Demand and supply play key roles on pricing and value realization. The 10% reduction in crop during 2020 due to the pandemic has increased the price realization up to 30%. Thus the timely supply of inputs and package of practices play a positive role on the sustainable productivity with quality, cost of production, and value realization for the sustenance of tea plantations. Any deviation/deficiency in the implementation of scheduled practices will impact corresponding crop loss, quality, value realization, and finally, the economy of the operations. The price fluctuation of tea in the international markets, despite the improved productivity with quality in many past years is a pointer for innovative reforms in marketing procedures, packaging, transportation, diversification, and value addition for the survival and prosperity of tea plantations globally.

CHAPTER TWO

Method of cultivation: propagation and multiplication of tea

Contents

2.1 Propagation methods

The tea plant can be propagated using both seeds and cuttings.

Plants raised from seeds are called seedlings. They show variations. In order to exploit the hybrid vigor, seed orchards are established utilizing elite clones and generative clones to combine the desirable traits in open conditions to improve the productivity.

Plants raised from cuttings are called vegetative propagation (VP plants). They are true to type and contain same qualities as that of their mother plants.

Grafting is another technique through which the vegetative parts of the plant is mechanically fixed on another plant, which grows like a single plant after successful union.

The cuttings are selected from mother bushes. Mother bushes are pruned and manured with extra care. The mother bushes are pruned twice a year, mature shoots are harvested, and cuttings are prepared and planted in nursery beds with overhead shade.

Since the starch content and auxin level play a vital role in successful rooting of cuttings, extra potash in the soil and fortnightly foliar of zinc sulfate are ensured.

In case of difficult-to-root cultivars, root enhancer formulations with indole butyric acid (IBA) in inert media are utilized to induce rooting.

Scientific Perspectives of Tea Plant Horticulture and Productivity
ISBN: 978-0-12-823444-0
https://doi.org/10.1016/B978-0-12-823444-0.00002-9

Care is taken to prepare the soil with adequate porosity, free from nematodes. Watering along with exposure to light in the morning and evening hours are routines to get the balanced root-shoot ratio. When cuttings grow satisfactorily with balanced root-shoot ratio in about six months, the plantlets are transferred to poly sleeves for maintenance in overhead light shade. Once the plant attains the height of 30—50 cm with brown stem indicating the maturity for planting, they are stocked in open nursery for hardening. (For further details refer to Tocklai Tea Research publications, Tea Encyclopedia, Hand Book of tea and Field Management, 2008; Hand books of tea of UPASITRI, Coonoor, Sri Lanka, Kericho, Kenya, depicting regional terran variations).

2.2 Raising plants from seeds

Similarly, seeds are also sown in raised beds, with lathe frames to minimize light incidence. Normally the nursery period is one year for both cuttings as well as seeds. Floater tests are carried out for the winter harvested seeds and graded, stored in moist sand/charcoal mix media, and periodically moistened, maintaining the humidity and preserving the viability.

2.3 Grafting

The grafted part is called the scion and the basal part on which it is grafted is called root stock. In grafted plants the scion provides the canopy of the bush maintaining all characters of the mother plant.

Root stock takes the responsibility of nourishing the scion-derived canopy by absorbing water and nutrients from the soil through strong root system and transporting them to the utilization site.

Clones suitable as scions include TV 1, TV 2, TV 14, TV 17, TV 21, S3A/1, S3A/3, and T3E/3. Clones suitable as rootstocks include TV 9, TV 18, TV 22, TV 23, TV 25, TV 26, and TV 29.

2.4 Maintenance of Seed Orchards (Biclonal seed baris)

Biclonal seed baris/orchards take advantage of the tap root in seedlings besides exploiting the hybrid vigor in natural breeding between the generative clones. Tocklai, TRA, Jorhat had been the pioneer in the novel crop improvement program, which amply benefited the tea industry.

The seed baris/orchards are provided with barriers and located with safe distance to avoid contamination. Beehives, wasps, midge flies, and a gentle breeze improve the pollination and fruit set. In order to encourage pollinators, the seed baris are seldom sprayed with insecticides. Ecofriendly, integrated methods and natural farming are used for control of bunch caterpillars, tea bugs and so forth. Natural control through birds is encouraged besides hand collection. The trees with new generative clones for new varieties, in principle, the China and Assam cultivars, hybrids are used. The quality and combinations, upkeep and maintenance, collection, and preservation storage are periodically monitored and advised by the qualified scientific inspectors to assure the purity and quality of the declared biclonal stocks (Figure 2.1).

After a reasonable period of bin storage in charcoal sand media, the seeds are raised in nursery beds to overcome the dormancy and suit the planting time, usually in April or October/Autumn or Spring. The seedlings are stacked outside and hardened for six months and looked after by watering, foliar application, and so forth by skilled workers. The plants are ready for planting in 12—15 months. After the reasonable period of exploitation of the combination, the bushes can be top-worked with bearers and pollinators by cleft grafting (Figures 2.2—2.4).

Figure 2.1 TS 520 (TV1 x TV20) fruits (capsules).

Figure 2.2 TS520 Seeds.

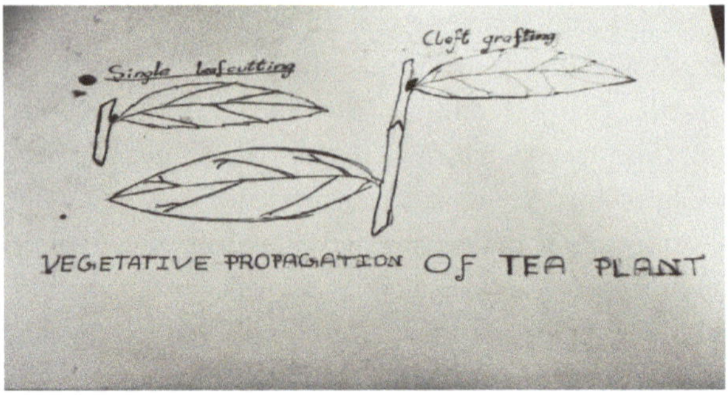

Figure 2.3 Single leaf cuttings and cleft grafting for propagation and conversion of seed orchards.

Figure 2.4 TS 520(TV1xTV120) a productive popular planting material for Assam.

2.5 Organic cultivation/natural farming of tea seed Orchards: Guidelines

1 **Site selection**: Elevated corner areas with water resources free from water inundation, a loamy soil with rich in humus will be ideal, just like any perennial horticultural crops located away from existing tea plantations with barriers to avoid cross breeding.

2 **Land preparations**: Level the ground, dig out drains to remove the excess water from the root zone border planted with shade trees as a barrier and ventilation, sheltering birds, and honey bees. Leguminous rehabilitation crops like *Tephrosia purpurea* can be grown for six months and plowed in for enriching the soil biosphere. Optimizing the soil pH, organic matter content enriches the fertility of soil with microbial population buildup.

Spacing: 5 × 3m spacing with 30 × 45 cm deep pits, allow for weathering and solarization. Fill up the pits with 2 kg of well decomposed cattle manure mixed with bioinputs like Trichoderma, Pseudomonas, Paecilomyces, Meterrhizium, and Phosphobacter, sp. Allow the soil to set for impaction and multiplication of the bioinputs. Raise intercrops like sunn hemp and daincha as mulch, and cover crops. These preparatory measures are taken during winter when the workforce is available.

3 **Planting Materials and Planting**: The cropper and pollinator cultivars are planted, alternated with each other and rammed up, provided with staking, and mulched, irrigating once a week.

4 **Cultivation practices-inputs/management**: The plants are manured with 3kg of cattle 2 kg of neem plus castor cake fortified with phosphor and k solubilizers per plant, twice a year in spring and autumn in the pit and incorporated, hand pulling the weeds if any and irrigated. Train the trees selectively, building up the frame and architecture, spray neem kernel extract (NKE), and herbal products to manage/control the bugs.

5 **Harvest and storage of seeds raised in nursery**: The trees start bearing during the third year. When the trees bloom in spring, spray nano micronutrients combined with amino acids for improving the fruit set and development. Similarly, the same spray can be repeated in autumn for fruit development and seed set. The capsules mature and shed the seeds in

winter. Thus the collected seeds are subjected to floater tests and the sinkers are stored in charcoal/sand media, periodically moistened, preserving the vitality and dormancy. Thus preserved seeds are raised in raised nursery beds, covered with thatch, and sprinkled with water to keep the moisture, facilitating germination. Once the seeds germinate, the mulches are removed, provided overhead shade, and raised. The seedlings are stacked outside and hardened after six months and looked after by watering, foliar application, and so forth by skilled workers. The plants are ready for planting in 12–15 months (Figures 2.5 to 2.7).

Figure 2.5 A view of typical nursery bed for VP of single leaf cuttings.

Figure 2.6 Plants in sleeves stacked for hardening.

Figure 2.7

CHAPTER THREE

Management of young tea plantation in field

Contents

1. Tea plant requirement: soil, plant, and water
2. Land: terrain and climate for growth and productivity
3. Nutrition: organic fertilizers at the time of planting and inorganic fertilizers follow up
4. Bringing up of young tea: training, pruning, harvesting (plucking), and postplanting care

3.1 Tea plant requirements

Soil: Sandy loam with pH ranging from 4.5 to 5.5, salt concentration (EC) less than 0.01 to 1.5 dSm-1, organic carbon 0.55%−2.5%, depending on the elevation and location from the equator, well-distributed rainfall throughout the year are some of the essential basics for tea cultivation.

1. **Assam**: Brahmaputra valley: Low elevation, sandy loam, pH ranging from four to six depending on the location (riverbed, valley, soil type, and ranges/elevation) intensive cropping and type of cultivars.
2. **Darjeeling**: Micaceous volcanic peat sandy loam soil with elevation 500−2500 m, organic carbon ranging from 3% to 10% with steep slopes/terrain/ranges, chinary cultivars, low productivity, and flavor high quality, without shade.

Scientific Perspectives of Tea Plant Horticulture and Productivity
ISBN: 978-0-12-823444-0
https://doi.org/10.1016/B978-0-12-823444-0.00003-0

Inherently the soil contains the metal strontium, which needs to be monitored in the made tea before export to European countries. Submontaneous Assam, Doars, and Darjeeling, Nepal area adjoining the mighty Himalayas, with plenty of water in the rivers mixed with rubbles and silt, combined with salubrious climate for tea (Gangra valley, Palampur, and HP) located in northern hemisphere.

3. **Nilgiris**: Southern India ranges, rolling hills, elevation ranging from 1000–2500 m MSL, lateritic peat soil with iron, calcium, pH ranging from 4.0 to 5.5 with high carbon 4%–8.5% surrounded by thick forest reserves, streams, rivers, and animal sanctuaries.
4. **Kenya**, Uganda, Tanzania, East and Central Africa: Volcanic, High elevations with high carbon up to 15%, optimum soil acidity (pH. 4–6) no shade, surrounded by reserve forests, located in Equatorial region with high rainfall; planted mainly by the Assam-Cambod hybrids of high productivity during early part of 19th century and started export in 1930. South Africa and Nigeria regions differ from East and Central Africa, Malawi, Rwanda, and Burundi are of later origin coming up mainly on hills up to 3000 m high, surrounded by rivers and well-distributed rainfall. The soil is of rich clay with humic top soil rich in organic matter (Baruah, 2015). Hence the package of practices and cultivar vary to suit the local terrains, climate and soil types. Pleasantly, tea plants adapt to local environment and biosphere gradually and perform to the inputs and well-designed package of practices (Owuor, 1999).
5. **Sri Lanka**: An island nation surrounded by seas, with rolling hills/ranges on the central land western side, soil, terrain, and vegetation are similar to South India (lateritic), elevation ranging from seashore (Galle) to 2500 m Nuvareliya and Uva valley, known for quality tea similar to Darjeeling. Sri Lanka manufactures mainly black tea—one of the best orthodox and CTC, exported to Europe and Gulf.

3.2 Plant/cultivar

Camellia sinensis with three subspecies, Assam, China, and Cambod, bred naturally and evolved very well as a heterogeneous cultivars with diploids ($2n = 30$), triploids ($3n = 45$) and polyploids. The natural combinations have contributed the evolution of versatile cultivars for many traits such as yield, quality, stress tolerance, adapting to different terrains, geographical locations, and biodiversity. The heterogeneous cultivars are the potential base line as breeding material for the future. Hence, tea has rich potential for breeding and selection of new varieties for any purpose of crop improvement, worthy of exploitation.

The life span of tea is more than 100 years. The care and attention given to the upkeep and maintenance of young tea reflect on longevity. The land preparation, pitting/trenching, inputs for filling pits, mulching, shade and cover crop, centering, formative prune, and frame formation (bush architecture) are some of the main operations in young tea.

Planting: Virgin soils are used for planting directly, while replanting fields are duly rehabilitated, soil acidity optimized with soil amendments, plowed, leveled, and rows are marked across the slopes in contours. Staggered double hedge is preferred to have a compact stand of around 15,000 plants per ha. Planting is done during fall and winter avoiding the monsoon. Tea research institutes and industry have developed and released planting materials. Proven cultivars suited to different agroclimatic conditions may be used to replace the old seed jats in the field.

3.3 Water requirements

The plantations are located wherever the annual distribution of rainfall is adequate, thus tea is treated as rain fed crop supplemented with ground water supply to meet the stress period in winter, nurseries and young tea in the formative stage. Water logging and flooding during monsoon weather are major problems in plateau and valleys of Assam/Bengal to maintain the optimum soil moisture: air; in the desired ratio 60:40 for sustainable productivity with quality; maintaining the health of the bushes. Though being a rain fed crop, water management is one of the serious problems in tea gardens of India. Being the plateau and low elevation, the plant material (cultivars) being broad-leafed Assam (betjan) jat, shade is obligatory in Assam to cut the incident light by 30%–40% with leguminous shade trees (Albizzias) besides wind barriers in the borders of the gardens to reduce the wind speed during monsoons. The cooler winter stress period is utilized with advantage for periodic pruning and maintenance and upkeep of the drains for effective drainage during monsoon. Thus, microclimate, environment, soil moisture, aeration, the biosphere of the soil and plant, play an important role in tea plant health and productivity with quality, impacting on the value realization and economy of corporate plantations.

3.4 Land: terrain and climate on tea growing and productivity

Tea (*Camellia sinensis*) is a versatile crop, growing in diverse climate, soil, and terrain with and without shade. Thus it is well adapted to local situations. It grows in Brahmaputra valley/plateau, with low elevation, making luxuriant growth, where ideal conditions of tea growth such as Brahmaputra valley/plateau, luxuriant climate with high humidity, sandy loamy acidic soil with adequate organic matter, excellent drainage providing optimum soil moisture, and environment (Willis, 2019).

Tea plant growth is moderated by leguminous regulated shade trees, ensuring the most productive range of soil and atmospheric (ambient) temperature and humidity. These are ideal situations for the growth and productivity of broad leaved Assam tea. The China tea prefers high hills with cooler environment and "micaceous peat soil" tolerating cold weather without shade. Though permanent shade trees not required for high elevations, ground cover with soft weeds and grasses help the microclimate, minimizing the soil erosion, and run off in the slopes, thereby preserving the top soil with humus and beneficial microbes favoring the health and productivity of soil and plants. Introduction of leguminous cover crops, shrubby leguminous green crops and semidwarf varieties of Albizzias and Acacias, and beneficial intercrops like mandarins, avocados, and beneficial proven herbs and spices are worth being explored with advantages rather than leaving barren and exposed. Need-based monitoring and regulation periodically will help the plantation biodiversity economically. These cover crops contribute conserving soil moisture aeration through root penetration and enrich the soil with root exudates, mutually helping symbiotically and contributing abundantly to the buildup of the favorable biosphere of the soil. The productivity of China tea is low, but the flavor and aroma are unique, fetching the highest price. The Cambod, on the other hand, a sturdy cultivar midway between the Assam and China in vigor, productivity, and quality serving as an adaptive cultivar for different terrains, climate, and a good filler in plantations. Besides these natural soils, microclimate, biosphere, and health of the plants, foliar application of organic molecules-micronutrients and biostimulants increases the photosynthesis of plant leaves accelerating growth reflecting on productivity with quality (Kempf, 2019).

The following are the distinct regions of global tea based on the location, soil, climate/terrain, type of tea made, and so forth: Assam, Bengal, Tripura, Bangladesh, Darjeeling, Nepal, Nuwareliya, Nilgiris, South India, Himachal

Pradesh, Sri Lanka, Thailand, Indonesia, Kulu valley, Kenya, Malawi, Tanzania, Uganda, Nigeria, South Africa, China, Japan, Australia, Russia (Georgia), Turkey, and Argentina.

3.5 Young tea management

Training, pruning and plucking (harvesting), and postplanting care.

Training of young tea to have proper bush architecture: The objective of raising young tea is to induce proper bush architecture through a series of pruning. This is obtained in three stages: (1) centering, decentering/lung pruning/debudding; (2) formative pruning; and (3) final frame formation pruning.

Centering is performed by knifing the leader terminal shoot and tipping the laterals, which occurs in early spring. Plants are ready for centering from April onwards. These two operations, centering and tipping in young tea, help to break the apical dominance, inducing the lateral to break from dormancy to grow while building up the canopy and bush architecture. Thus the apical dominance governed by auxin is broken through the cytokinin produced from the roots and the gibberellins produced from the new leaves and lateral growth is induced. The scientific knowledge of phytohormones and their interaction and balance has been approached by the pioneers breaking the apical dominance in young tea management (Tocklai, Encyclopedia, Handbook and serials, 1950–1988, DN Baruah, 1989). There are clones and seedlings where difficulties in getting adequate lateral sprouts are encountered.

In order to break the apical dominance, various practices, such as pegging, decentering, lung/thumb pruning, debudding, and the use of growth retardants were evaluated, and packages/practices developed and adopted. Nevertheless, pegging is the most economical and ecofriendly practice for conserving the dry matter, although it takes time and is more labor-intensive (Manivel, 1987 Adv. Bulletin on lateral promotion in young tea, 1987).

Shoot apex dominates the lateral buds in growth, forcing them in to quiescence/dormancy in perennial plants like tea. This phenomenon is governed by the auxin, indole acetic acid (IAA), and a concentration of its balance with growth retardants like abscisic acid and ethylene. A high level of auxin in the tips triggers synthesis of ethylene, which can act synergistically based on the concentration and weather.

Apical dominance is broken by removing the tip, breaking the stem, pegging, debudding the main stem, and altering the balance between the promoters and inhibitors by using plant growth promoters like gibberellins. In weak plants and seedlings, induction of laterals is a bit difficult and erratic due to genetic complexity, where all these measures are used to induce lateral bud break, breaking the apical dominance.

Thus the first step of building the desirable bush architecture has been commenced by training the bush to facilitate the future management operations, especially the harvesting of tender shoots periodically.

3.6 Tipping and plucking

Tipping the shoots periodically induces branching and spread of the frame, building up the top-hamper/canopy. Thus by raising the tipping height (35, 40, 45, and 55 cm) yearly, the plucking table is formed at 90 cm, a convenient height for the women workers to harvest the crop. Formative pruning is done in the third or fourth year depending on the growth, when final cleaning, desnagging, opening the center, and fine tuning the frame occurs. Thus the desired bush architecture balancing the wood and mature maintenance leaves (canopy) ensured facilitating smooth sap flow, photosynthesis, and consequently sustainable productivity with quality. Weekly selective standard plucking, maintaining at least 65% fines and 34 rounds per annum, should be ensured.

3.7 Frame formation prune

Lower frame with an adequate number of primary/secondary branches, wider angle between the branches to sustain sap flow, three to five higher shoot intensity with pubescent tips, and sufficient foliage are preferred. Whatever the methods followed, selective knifing is inevitable, and dry matter can be saved. Earlier ground cover is advantageous in depressing weed growth and establishment. In the process, spacing between bushes has been reduced increasing the number of bushes per ha. By refining these parameters, the harvest index has increased from 7.5% to 11%. Thus, importance of bestowing more attention in establishing and managing young tea fields have been amply demonstrated.

A three-year-old TV23 (a standard yield clone for Assam, Bengal) double hedge staggered planting with high density succulent shoots ready for harvest. With proper management utilizing the available technology, the clone has potential for 10,000 kg per Ha from the fifth year of planting.

Plates. Illustrations on spacing, height of frame, shoot density, and bush frame (TV 23 field).

3.8 Schedule of operations for bringing up young tea

This operations is influenced by agro-climatic conditions and the elevation.

1. 0 year: October—December: plant tea and allow to grow;
2. +1 year: Late February to mid-April/June to July: Decenter at 20 cm, and tip at 65 cm above ground. Head back strong central branches and remove crossings. End October: step up by leaf.
3. 2 Year: Whole year: Prune and pluck to janam.
4. 3 Year: End January to early February; end of October to first FFP at 35—40 cm, tip at 65—70 cm and step up one leaf if required.
5. 4 Year: Whole year Prune and pluck to janam.
6. 5 Year: End January to early February; Final FFP at 40—45 cm. Head back thick central branches (if any). Tip at 70 cm and pluck.
   Photo showing a YT field with high density of shoots TV 22/23.

3.9 Postplanting care

Postplanting care includes shade, nutrition, mulching raising cover, green crop, and plant protection measures including weed control measures.

1. **Mulching and cover crops**: Mulching the soil with dry leaves and humus generates CO_2 in the bush ambient area. The increased CO_2 concentration increases the photosynthetic efficiency of the leaves; contributing additional saccharides and amino acids (assimilates) that will reflect on the productivity with quality through metabolic acceleration. Linear Response in Photosynthetic rate had been found up to 500 ppm of CO_2 under steady-state conditions of intact mature tea leaves. Beyond this limit retardation sets in through stomatal closure. Similarly, humidity in the canopy ambience enhances the photosynthesis of leaves through stomatal pores, opening gas exchange. Transpiration through stomates cools the temperature of leaves sustaining the photosynthesis rate. Thus, mulching and cover crops in the fields contribute for enhanced productivity up to certain limits.
   The biosphere of soil and climate have a bearing on the plant on the whole, impacting the health and productivity of plants.The plant parts below and above soil include the roots and canopy(frame and shoots), which are mutually dependent and coordinate the plant metabolism reflecting on the health and productivity. The nature of the biosphere again depends on the soil type, acidity, organic matter content, aeration, and moisture content contributing to the spread of the root system,

buildup of beneficial microbes, and absorption of nutrients regulating the metabolic functions of the canopy. The pH and organic matter (humus) content in turn play a role on regulating the retention of nutrients and enzyme activity as buffer nourishing the microbial population and enzyme activities impacting the growth and productivity of the plants. Mulching and cover crops contribute to this process, by adding organic carbon, moisture and aeration moderating the growth and multiplication of microbes, enzymic reactions, and thereby the absorption of nutrients stabilizing the metabolic functions of the canopy sustaining the growth and productivity with quality. The leguminous nature of shade and cover crops also help in adding additional nitrogen source through nodules and root exudation. The enhanced CO_2 concentration in the canopy region increases the photosynthetic rates of leaves (Manivel and Hussain, 1987), contributing to increased productivity without any additional inputs. The resultant effects of enhanced humidity, shade, weed control, and congenial biosphere due to mulching and cover crops are significant and deserve to be appreciated and implemented (Willis, 2019). Strong winds also affect the plants' metabolic processes, such as transpiration, gas exchange, pressure on stomate opening, and ultimately suppressing the photosynthesis. Therefore it is imperative to provide good wind barriers, especially on the windward direction, to arrest the wind speed during a monsoon and also to reduce soil wash, erosion, and demarcating the boundaries besides protection from trespassing wild animals.

2. **Shade-stand and regulation**: The shade trees in plantations regulate the ambient temperature, light intensity besides humidity, sustaining the photosynthesis and assimilation. Tea plants relish moderate shade and benefit against high intensity and temperature, impact of heavy winds, and so forth. Thirty to 40% of full sunlight and 25°C–35°C leaf temperature are ideal for photosynthetic efficiency. Higher range beyond this will trigger respiration—both photo as well as dark respiration—burning the sugars. Over shading reduces the sunlight incidence, the leaf temperature as well lowering the photosynthetic efficiency. Therefore, regulation of shade through lopping has to be resorted to optimize the shade and light for sustaining the Pn.12–15 $mg/dm^2/hr$ of the tea leaves. Because the shade trees are leguminous, the roots help fix nitrogen. Leaf fall during winter also adds up nutrients, enriching the soil. Thus, optimum shade with seasonal regulation will impact additional crop up to 10%–15%.

3. **Moisture Stress**: Soil moisture and aeration play an important role in sustaining the metabolic functions and productivity. Both drought and flooding affect the photosynthesis of the canopy leaves. Aeration helps the beneficial microbes and enzymes to sustain the absorption of nutrients by roots and assimilate in the canopy leaves. Optimization of soil moisture and aeration at 60:40 ratio through drainage and irrigation, though cost intensive, fetch at least 10%–15% extra crop.
4. **Pests and Diseases**: Both pests and diseases weaken the plants affecting the productivity up to 12%–15%. Pests like helopeltis suck the sap from the leaves, injecting toxins that burn the tissues. Semiloopers (*Hyposidra talaka*) completely defoliate causing maximum damage. Jassids, thrips, and aphids are mild pests sucking the juice of leaves. All these are affecting the leaves in the canopy decreasing photosynthesis and metabolism.
   Diseases like red rust (*Cephaleurus parasiticus*), blister blight (*Exobasidium vexans*), and black rot (*Corticium*) reduce the leaf area thereby affecting photosynthesis and blocking the sap flow, causing defoliation and dieback. Root disease like ustulina block the root canal with mycelium, ultimately decomposing the roots. Slow decline takes place, and ultimately death causes vacancies in patches.
5. **Chemicals-Herbicides, pesticides, fungicides, and residues of sprays**: The contact herbicides like paraquat scorch and burn the tissues, killing the leaves, whereas the systemic herbicides like glyphosate gets translocated killing the apical buds. The chemicals in other mature leaves at toxic levels interfere with the metabolism of the leaves, causing mottling symptoms that ultimately lead to scorching and defoliation.
   The glyphosate in roots and leaves trigger the respiration burning up the reserves, debilitating and killing the bushes. These spray chemicals also kill useful predators and parasites in the canopy besides killing the microbes in the soil, depleting their population. The damage thus is permanent and costly. Chemical effects persist for long durations, causing yield reduction up to 7%–10%, ultimately casualties/vacancies. The chemical residues affect the quality of the produce.
6. **Soil and soil nutrition management in young tea**: Sandy loam is the most preferred type for tea. Too clayey or too porous sandy soils are avoided. However, any available soil can be corrected and amended if all other parameters are conducive and favorable.
   Acidic soils (pH 4.5–5.5) are cherished by tea bushes. When the soil is too acidic or alkaline, many micronutrients become unavailable, causing

deficiencies impairing many metabolic activities, with ultimately weak and low productivity.

Deficiency of zinc, magnesium, manganese, molybdenum, and boron elements will have a serious impact on photosynthesis. Production of quality constituents like ascorbic acid, kinase enzyme activity, and oxidative enzyme activity are hampered reflecting on the productivity and quality of the produce, besides making the bushes unhealthy. Humic acid content and microbial population are directly correlated with aeration of the soil, and feeder root proliferation, thereby absorption of nutrients.

The population of Acetobacters, Aspergilus sp, nitrogen-fixing bacteria, Clostridium for legumes, Pseudomonas, and the beneficial fungi like Trichoderma, Entomopathogens like Paecilomyces, Metarrhizium depends on the humus content. The enzyme activity of nitrate reductase and others depends on the potassium humate, which is a good buffer and the food for the microbes, facilitating their function through aeration (Harold Willis, Kempf, 2019).

7. **Organic nutrition for young tea at the time of planting**: Depending on the soil, cultivar, and bush population, the inputs of planting pits vary. However, based on the technology presently available, protocol for preparing the pit mix for planting pits is given in Table 3.1.

   These inputs are designed to protect the plants from pathogens in the soil, nematodes, and termites, which induce root growth through aeration and slow release of the P required for root growth, besides supplying the nutrients required for general growth and ground cover of the plant in initial establishing stage.

Table 3.1 Composition of PIT mix for planting.

S.No.	Composition	Quantity
1	Compost/cattle manure	2 kg
2	Azospirillum	10 g
3	Phophobacter	10 g
4	Trichoderma	5 g
5	Paeciliomyces	5 g
6	Pseudomonas	5 g
7	Metarrhizium	5 g
8	Rock phosphate	100 g

The above ingredients to be mixed thoroughly and filled up the pit along with excavated top soil to the brim and allow to set before planting.

*All hard chemicals like methyl bromide, mercuric chloride, and chlorpyriphos are prohibited to conserve the microbial population and health of the Root system.

8. **Inorganic Nutrition for young tea**: Nutrition in the form of inorganic fertilizers in graded dose, gradually raised from training, to build up of frame until formative pruning occurs in the fourth or fifth year. As regular production commences from the fifth year of planting, a standard dose of NPK given in split doses depending on the cropping season, based on the productivity, and rationalized for sustained productivity (Tocklai Tea Encyclopedia, 1950–1988 pp 647).
   As a policy NPK is given in 2:1:2 ratio commencing from 100 to 150 kg based on various factors like cultivar, soil, age of the bushes, population, productivity, pruning, and so forth. Due consideration and compensation/weightage is given to the type of soil, such as clayey, sandy, or sandy loam, and status like pH, organic matter content, and microbial population. As the plantations get old and exploited with high population and staggered pruning, micronutrients deficiencies do appear. It would be better to get it corrected with foliar spray so that productivity is sustained, maintaining the bush health (Willis, 2015).
   During light limited situations like August month in Assam, and Doars due to overcast heavy clouds, triacontanol plus zinc foliar was found helpful in restoring/sustaining crop through stabilizing photosynthesis in subdued light conditions.

3.10 Mature tea plantation management

This could be termed more as shades regulation management on account of the contribution extended by the shades trees. They also preserve soft weeds that support soil erosion and run off. Shade regulation is done by permanent shade as well temporary shade. Shade is regulated to have an optimum level of photosynthesis. It is 12–15 mg/dm/hr in Assam conditions. Infect, the tree wealth coordinates the principal moderators such as moisture, pH, and electrical conductivity (EC) helps in optimizing the vulnerable parameters that ensure plant growth and productivity. The organic matter build up helps in increasing photosynthesis. Growing cover crops as green mulch also increases the organic matter/humic acid. Periodic soil amendments is a remedial measure against soil pH fluctuations. Need-based foliar applications of micronized sea weed, microbial extracts, and organic growth promoters with nanomicronutrients increases the photosynthesis, impacting the productivity and quality of the plants.

A scientific method of drainage should be designed so as to keep soil moisture and air ratio at 60:40 lm11321.

CHAPTER FOUR

Mature tea (soil, water and shade) management

Contents

4.1 Introduction

Soil, plant, and water are the main resources that govern the productivity, quality of the product, and value realization. The fourth aspect, light, is freely gifted by nature, which has also been regulated judiciously to reap the full benefit. Shade trees and wind barriers play a major role in this process. Tea loves moderate shade to tide over the rigor of summer heat and bright sunshine. While the full sunshine gives about 1600 micro-Einstein (μE), tea leaf photosynthesis is almost saturated around 300−400 μE. Hence leguminous shade trees in plantations of plains and low elevations benefits the productivity as well as the quality of the produce (refer to the graph on linear response of Pn rate of intact leaves to increasing the light intensity) (Figure 4.1).

Scientific Perspectives of Tea Plant Horticulture and Productivity
ISBN: 978-0-12-823444-0
https://doi.org/10.1016/B978-0-12-823444-0.00004-2

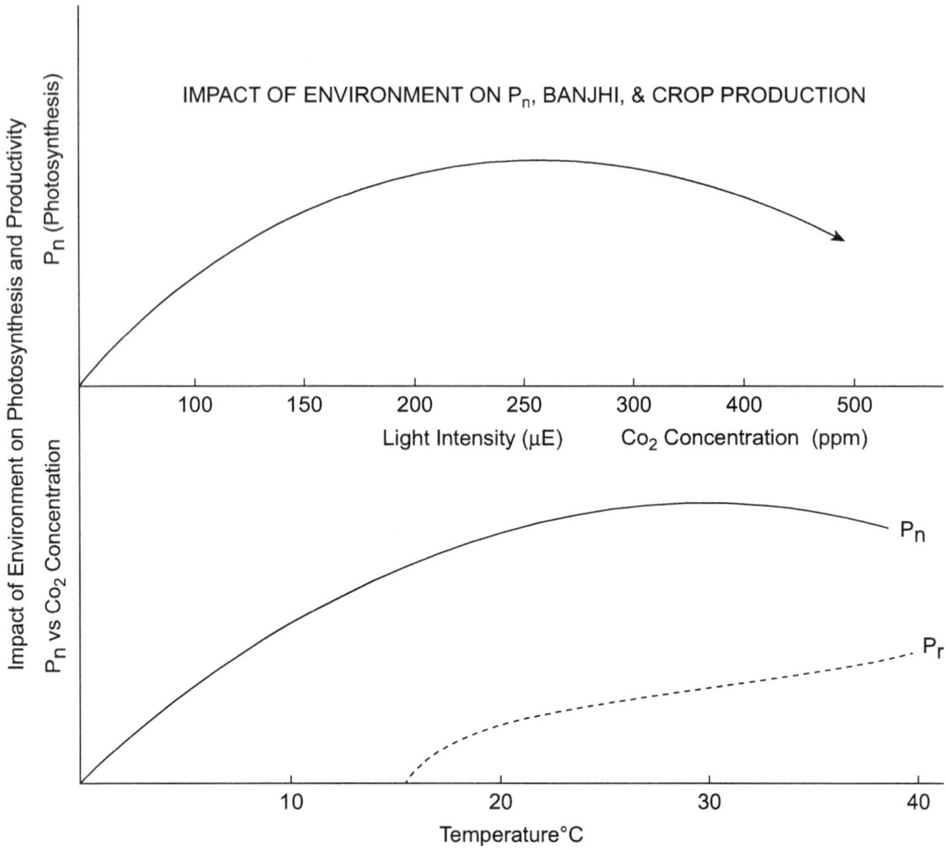

Figure 4.1 Impact of environment on P_n, Banjhi, and crop production.

In principle, leguminous shades are preferred. Mixed interplanted or alternate rows, 10–15 m apart, is recommended. In order to optimize the light intensity shade regulation is done prior to onset of monsoon.

To maintain the health of shade trees, lime-slurry banding of the collar/trunk and soil application of dolomite at 2 kg/tree in the winter is recommended, since shade trees require higher pH than the tea bushes, protecting the bark from sun scorch and gummosis.

Shade trees serve as moderator of light intensity besides contributing about 20 kg of nitrogen per hectare, through leaf fall and root fixation of N. Leguminous plants are preferred as permanent shade while other species like Gora Neem (*Melia azadirach*) can be used as temporary shade with advantage.

4.2 Soil

Soil is a living media for microbes, earth worms, and other beneficial flora and fauna. The soil conditions should be optimized to support these flora and fauna. Porosity of the soil, soil acidity (pH), electrical conductivity (EC), organic matter content (humic acid), and moisture content are the principal moderators ensuring the plant growth and productivity (Willis, 2019). The importance of a healthy soil with its structure and constituents, biosphere, and plant health for natural farming for sustainable productivity with quality have been amply explained.

Periodic soil amendment is a remedial measure to optimize these vulnerable parameters in order to sustain the productivity with quality. The optimum range of soil pH 4.5–5.5 is ensured by periodic checking and correcting the pH either through dolomitic lime (calcium magnesium carbonate) or iron pyrites, magnesites, sulfur, or humic acid (potassium humate) depending on the pH and salt concentration of the soil. Growing of rehabilitation crops like Guatemala grasses (*Tripsacum laxum*), *Citronella javanica*, legumes Mimosa (*Mimosa invisa*) and cover green crops *Tephrosia purpurea/ anagyroids, Sesbania speciosa, Sesbania aculiata,* and *Crotalaria juncea* are some the species cover green crops grown in low-lying riverbed area where the pH is marginally high around 6.0 due to sedimentation of soil wash, help amending the soil acidity biologically by adding organic matter while enhancing nitrogen fixing. Besides, biomass producers like *Glyricidia maculate* and Soo babul (*Leucaena glauca*) can be planted in the borders/out skirts wind barriers and temporary shade trees. These trees are lopped periodically, incorporated for the young and mature tea in the staggered furrows/trenches, augmenting the OM, humic acid as mulch, enriching nitrogen, build-up of microbes enhancing the porosity, and the root proliferation affecting better absorption of added nutrients. As a result of these OM build-up and mulching, the ambient CO_2 level goes up, increasing the photosynthetic efficiency of maintenance leaves in the canopy reflecting on the growth and productivity of young and mature tea. Thus these biomethods of soil reclamation can be done with advantage instead of going for chemical amendments wherever feasible. Nevertheless, added calcium through dolomitic lime helps the plants with secondary nutrients like calcium and magnesium. The current strategy is to skip rehabilitation with Guatemala grass because the two-year gestation period is expensive and depth of root system is concentrated in upper soil up to 10–15 cm depth is aimed to be fertile, adopting close planting of higher population for intensive cultivation and uprooted after a live span of 25–30 years instead of keeping perennial. Considering the economics and viability of marginal tea plantations, this practice of shorter reclamation period, replanting with adequately rationalized inputs, is considered as one of the alternatives for diversification. Growing the cover, green crops add up nitrogen by enriching the soil with root exudates creating a favorable biosphere for growth and establishment of young tea. Thus, the refined procedure of replanting avoids deep tillage besides reducing the gestation period preferred by planters for the sake of the economy.

A congenial ratio (60:40) of moisture and air in the soil is asserted in natural farming and ecofriendly integrated good agricultural practices (GAP) to ensure the sustainable productivity with quality and economics. The soil moisture versus the aeration of media is ensured and regulated through effective drainage and outfall of excess water during monsoon, besides enriching the organic matter content of soil, build-up of beneficial microbes for promoting the root proliferation and utilization of the available nutrients. Any deficiencies in these optimum parameters of the soil and plant are timely ameliorated and corrected with appropriate inputs to restore the health and productivity (Tocklai Encyclopedia of tea 1950–88, and UPASI Hand book of tea 1971). Enzymes are activated for the conversion of nitrogen into nitrates facilitating absorption. Depth of root system of plants depend on the soil depth and groundwater table. The root exudates and the microbial population, including the humic acid, soil moisture, and aeration play a vital role on absorption and metabolism of nutrients and water through roots to sustain and growth of the plants for sustainable productivity. Need-based foliar applications of micronized organic nutrients including biostimulants/growth regulators increases the photosynthesis and growth reflecting on the productivity and quality. Judicious, timely application of these micronized bioinputs increase the growth, impacting on the productivity with quality for sustainable productivity preserving the health of the plants (Kempf, 2019).

4.3 Water management

Soil moisture plays an important role on the health and productivity of the tea plants. Both excess and low moisture content affect the metabolism and the productivity, therefore the moisture content should be optimized to sustain the productivity. In general 60:40 ratio of moisture and air is preferred for capitalizing the metabolism, productivity, tolerance to stresses, and therefore health and productivity.

Tea being a monsoon-dependent crop exposed to both excess (water logging) and scarcity of soil moisture (drought) influencing the crop distribution peaks and troughs, exposing to various stresses like water logging, drought and incidence of pests and diseases, warranting timely prophylactic and palliative measures (Figure 4.2).

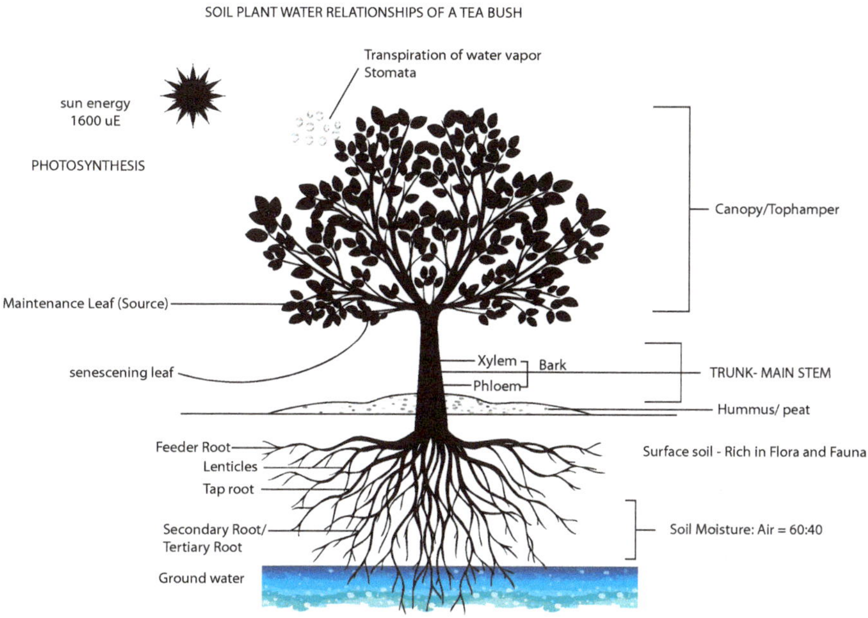

Figure 4.2 Soil plant water relationships of a tea bush.

Provisions of adequate drains, sprinkler/drip irrigation are some of the measures augmenting the soil moisture to prevent the water stress. Building up the organic matter content thereby the humic acid as a buffer and food for the microbes and enzymes, heavy mulch preserving the microclimate (biosphere), conserving the soil moisture, feeder-root proliferation absorbing the nutrients are some of the known benefits that accrue through proper management of soil moisture. Judicious handling of soil, water, nutrients management, effective timely harvest of crop, processing and aggressive marketing will ensure the productivity with quality for better value realization (sustainable productivity). Drainage should be designed and managed in such a way, water table is maintained below the root zone. It should not submerge the roots beyond 24 h.

4.4 Shade trees in tea plantations

Shade tree and its management is very important to regulate the physiological metabolism of tea plants, soil moisture etc. Average photosynthetic rate in Assam conditions vary between 12 and 15 mg/dm/hr. This can be manipulated to the best advantage, providing optimum/ideal conditions in the fields. Unique example of enhanced efficiency and productivity is of Kenyan tea plantations. The plantations are located in equatorial region, without shade, around 1500 m elevation, volcanic soil with high carbon (12%–14%) producing up to 5000 kg per ha.

The cultivar/plant material are of Cambod-Assam hybrid originated from Tocklai, Jorhat Assam. The bush capacity is around 10,000 kg/ha. Thus, the productivity of Indian tea plantations can be easily raised to the Kenyan level by optimizing the field situations, cherished by the tea bushes. The soil and water management besides the biosphere of the plants are the pointers for the future to protect the environment. Awareness have come for restoring the ecofriendly cultivation of tea with integrated nutrition management (INM), integrated pest management (IPM), liberal use of organic bioinputs, sustainable productivity with quality protecting the environment and water resources, avoiding hard chemicals either for plant protection or weed management.

Sri Lanka uses a three-tier vegetation of shade system: *Albizzia mollucana*, *Glyricidia maculata*, and *Erythrina indica* (Dadaps), periodically trimmed and lopped adding up a valuable rich organic matter/humus to the soil with advantage.

In south India, Silver oak (*Grevillea robusta*) is used as the shade tree which also serves as the wind barrier. It serves as the timber also. The planters hesitate to use the Dadaps as shade in tea assuming that the cation exchange capacity of this plant competes with tea. However, there is no data to prove this aspect. Dadaps serve as a good shade for coffee, cardamom, and other spices in Southern India.

Plantations of high elevations like Darjeeling and Nuwareliya may not require shade. However, leguminous shrubs like Sesbania, Crotolaria, Mimosops, and herbs like *Arachis alata* can be used with advantage besides preserving the soft weeds like Oxalis, Boreria, *Centella asiatica*, algae, lichens, and so forth for controlling soil erosion and run off, protecting the soil health and biosphere. Staggered trenches along contours planted with either citronella or Vettiver worth arresting the soil wash/erosion and moderating the microclimate. It is strongly recommended though, to provide shade and wind barrier in lower elevations up to 1000 m of these regions to tide over, storms, and arrest landslides, besides contributing to the build-up of organic matter, helping the microbial population build up, in addition to creating and maintaining the desired microclimate sustaining high rate of photosynthesis of maintenance canopy. Impact of low temperature effects due to elevation is nullified. The shade status can be regulated depending on the season by adjusting the spacing, mixing of species, and regulating through lopping which will add up the valuable OM/humus instead of no shade. Higher mixed shade with regulation has proven better than no shade in any tea plantation. Especially under the unpredictable climatic changes prevailing globally.

As for Assam, Doars, and similar terrain and climatic conditions of Central Africa (Kenya, Uganda, Tanzania) developed by pioneers/entrepreneurs, the shade patterns and maintenance have been standardized and holding with need-based refinement and modifications (Tocklai, TRA: Tea Encyclopedia,1950–88, Handbook of tea, NTRF Kericho, Kenya, 2002) are holding and practiced with advantage. The salient points on shade requirement, maintenance, regulation, and their role on sustainable productivity with quality are shown below:

1. Permanent Shade: *Albizzia chinensis*, odoratissima, *Acacia lenticularis*, *Derris robusta*, Pavonina, *Dalbergia sisoo*, mixed, spacing 10 × 5 m, uprooted the alternate trees after 5 years maintaining 10 × 10 (40′×40′)
2. Temporary shade: Indigofera teysmanii, Melia azadiracht/dubia (Gora Neem) Glyricidia maculata, Leucaena glauca (Soo Babul)
   Hedge grown in alternate rows, lopped, and incorporated periodically
3. Cover crops: *Tephrosia anagyroids*, purpuria, *Crotaleria juncea*, *Sesbania speciosa*, *Sesbania aculiata*, and *Cajanus cajan* (Arhar)
   Sown in furrows of alternate hedges, lopped, and incorporated, uprooted after 2–3 years
4. Green mulch crops: *Arachis elata*, herbs, soft weeds sheltering predators, preserving soil moisture, microclimate, self-subsisting; symbiotic with micoryza and root exudation

4.5 Role and benefits of these green vegetation

They contribute and provide:

1. Shade to moderate the leaf/ambient temperature enhancing the photosynthesis, thereby inducing growth and productivity.
2. Adds nitrogen, other minerals through nodules and leaf fall.
3. Help soil porosity, aeration, and build up microbial population and humic acid.
4. Absorb and feed the nutrients from deeper layers to feeder roots of tea.
5. Enrich the soil through exudates, driving many enzymic reactions.
6. Preserve the colloidal property of the soil, diluting the salt build up/hard pan, promoting absorption of inputs, moderating the soil acidity as buffer and amendments.

7. Trees like Grevillea, Sisoo, Red sanders, *Cassia siamea* serve as A-class timber is handy at emergencies.
8. Wind barriers especially in the windward boundary, check the speed of winds and minimize the damage of hail storms.
9. Fast growing vigorous trees like Spathodia, Boerhavia(Jahrul), Eucalyptus citriodora, Bamboo, Delanix/poinciana regia (Flame of Forest) *Albizzia mollucana*, odoratissima, Narayna (Marudhu Terminalia sp) in the banks of ponds, streams, and reservoirs pump the excess water by transpiration reducing the water table during monsoons in tea fields.
10. Trees like Pongamia glabra(Karanji), Litchi, Jamun, Jack fruit (*Artocarpus integrifolia*/incisa) Ficus, *Terminalia catappa*, and *Bassia latifolia* invite birds harboring the area, clean up the pernicious caterpillars, insects protecting both young and mature tea economically conserving the biodiversity.

Thus provision of shade trees, cover crops, green mulch, wind barrier, and fuel agro-forestry play a significant role in tea plantations. Judicious diversification/intercropping with suitable horticultural crops in tea plantations are worth consideration.

(Manivel. L Personal communication, 2019).

A photo of a tea garden in Southern India with full shade complements.

A organic tea garden with good shade and canopy, high productivity with quality, and value realization, in Assam.

4.6 Water conservation and management

Judicious, need-based rationalized economics of our soil, plant, water, and nutrients management are the domains of agriculture, horticulture, and plantation crops. Water, being an important inputs, deciding the growth, productivity, and metabolism of plants, it deserves consideration by all scientific researchers and managers for achieving sustainable growth, productivity with quality for the satisfaction of consumers, especially for the perennial crop-like tea, in the changed climatic conditions of today. Tea being a perennial rainfed crop, quality water is required for foliar application, fertigation, and processing-humidification and cleaning the fermenting floors and machineries. The nurseries with young tea require cleaner water. Of late the erratic rainfall distribution; climate change, diversification, advanced refined cultivation practices, drip irrigation/fertigation are required in plantations for stress amelioration. The need for conservation and management of water in tea plantations are paramount for achieving sustainable productivity with quality, maintaining the health and longevity of tea plantations.

Much efforts in water and fertilizer resource management are directed toward optimizing their use for sustainable productivity with quality, adopting GAP, minimizing the adverse impact on soil, plant and environmental impact, and productivity (Chatterjee, 2018). Similarly the rainfall intensity and distribution varies due climatic changes in different regions of tea in Assam, NE, warranting conservation and usage measures/reliable package for the sustainable productivity, minimizing the impact of flood/water logging and drought, in Assam and Arunachal Pradesh in particular (Sarma, 2018).

The free smooth flow of rain water in the streams/rivers from the garden out lets should be ensured to retain the water table in the gardens below the root zone avoiding water stagnation/water logging more than 24 h for preserving the roots. Discharges of any effluents, contaminants from factories, refineries, brick kilns, etc., should be prevented ensuring cleaner environment benefiting the flora, fauna, and biodiversity. Soil-binding grasses, shrubs, and hardy quick growing trees, and so forth can be raised in the boundaries of rivers, streams, ponds, and reservoirs, preventing erosion and soil wash while conserving the water, soil, and environment.

Principal causes/reasons for the flood and drought in Assam North East:

1. The mighty river Brahmaputra changes course, inundating the valley every year. Although its silting effect improves the soil, water logging affects the peak season crop during monsoon, predisposing the bushes for root diseases. The tributaries also erode the soil, carry enough silt and shallowing the depth and overflow. The unmaintained dikes, along the upstreams and encroachments in the river banks, also contribute for the soil erosion, flood, and seasonal water logging, and every year a recurring phenomenon causes havoc to the plantations and rural population. The causes should be rectified. The outfall problem from the gardens should be solved by the garden while the tributaries' back flow problem should be rectified by desilting, deepening, clearing jungle growth, sand dunes preventing the entry in to the big river has to be attended and restore flow every year by the PWD, state department. Attending these two operations by the gardens, in coordination with the state and federal governments every year will help to reduce the damage of the gardens in Assam, increasing health of the bushes, productivity, quality, and price realization, which are aspired for by the plantations and people of Assam along with the entire North East for improving livelihood and prosperity.
2. Because of the high water table in fields, the root zone of the tea bushes and shade trees are shallow, vulnerable to storms during monsoon times. Consequently, plants go through severe stress of drought with concomitant inability to exploit the deeper layers of soil for nutrients when needed, aggravated by drought-induced pests and diseases.

3. Rainwater conservation measures essential to conserve the rain water during a monsoon that can be used for field and factory operations where clean water is required. The existing infrastructures need to be expanded and strengthened for improved production with quality. Both fields, and processing will be benefited in the long term. During winter the plants go through soil moisture stress, which can be minimized by fertigation and preserving the health of the plants, which will certainly boost the early crop after winter.
4. Deeper root system with augmented absorption of water and nutrients will minimize the damage of both roots and plant systems with improved health of the plants, which will go through drought free from pests and diseases, paving the way for early improved crop, with quality, reflecting on favorable crop distribution and economy and cost saving in plant protection. Already these measures are in place but inadequate due to poor maintenance, encroachments, and blockages. The water resources in Assam, Arunachal, deserve to be surveyed again, and blockages and encroachments are removed, ensuring the free-flow of water during monsoons without inundating the plantations/fields. This will reduce the flood havoc during a monsoon, minimizing the soil wash, erosion of the plantations and the low-lying rural area, enriching the soils, and preserving the biodynamics of the region maintaining the ecological balance and environment. Besides these measures of water conservation damage control in plantations, enforcing permanent long term measures of developing the Brahmaputra River and horticultural wealth of the regions will benefit the population and the nation. This strategic, joint venture national project of improving the horticultural, forestry wealth, and transportation through inland water transportation are envisaged to have salutary effect on the economy of North East as a tourism spot with economic advantages.
   1. The course of river Brahmaputra should be stabilized by periodic dredging, deepening, raising the bunds, and forestations with tall growing sturdy timber trees up to 5–10 km in the boundary on both north and south banks.
   2. The boundary of the riverbed should be strengthened by raising lemon grass, citronella, Pongamia, Eucalyptus, Naraina (*Terminalia* Sp) inclusive at close spacing with deeper roots serving as barrier against flooding the area, effecting free-flow of water in the downstream.

3. Horticultural crops like fruits and vegetables can be organized on a cooperative basis for precision farming. The harvested produce should be supplied to the warehouses and processing units and entrusted for marketing.
4. River sand is very much in demand for real estate/realties. Supply can be initiated by ferrying the dredged sand to the builders, earning good revenue for the exchequer to augment the expenses.
5. Many tea gardens of the tea industry, including the Assam Tea Corporation, about 17–20 numbers are in crisis/distress, due to various deficiencies, including financial crisis and maintaining the gardens with lakhs of depending work force settled in gardens for five generations. Restoration and revival of the plantations is a priority area under consideration. Implementation of innovative reforms, package of practices, mechanization, natural/precision farming, soil reclamation, consolidation, improved manufacturing, diversification, value addition, and marketing skill, etc., are envisaged for early revival and restoration of the plantations with minimum damage, adopting the technology and resources available in the country. The labor welfare measures ensuring their settlement creating avenues for settlements and employment opportunities for the lakhs of able-bodied, trained, skilled dedicated work force on priority with justification. The required avenue and opportunities are to be created for enabling the millions for joining the mainstream is the top priority of the democratically, elected, performing government under the leadership of the Prime Minister of India. A few lines on the course of action on priority with timeframe are narrated below for consideration, as a well wisher in the interest of the prestigious tea industry and the nation.
   1. The surplus workers can be settled in the bank of great river Brahmaputra valley of Assam with their consent where afforestation and horticultural projects are envisaged with processing, and marketing facilities proposed to be created for the development of the North East.
   2. Skill development training centers created for the willing able-bodied boys and girls to enable them to join and work in fields, warehouses, and processing factories including marketing through surface water transportation envisaged.
   3. Cultivable land of about 2–3 ha per family could be allotted on lease for cultivating horticultural crops, fishing, etc. for their livelihood. The organic produce can be procured by the

cooperative body (production co.) and warehouse, feeding the processing units established all over Assam, Majuli Island, Gholahat, Tezpur Guwahati, and marketed.

4. The North East has fertile soil and a climate suitable for various flora and fauna. Horticultural crops can be judiciously exploited, harnessing solar energy and environmentally friendly packaging of practices enhancing demand for cleaner value-added products globally for the welfare and prosperity of the nation. The new settlements created by the Government of India in coordination with the states of North East can be focused as an example for the ecofriendly development of the people below the poverty line as a model ethnic village of tea garden workers in Assam, originally recruited and brought by the pioneers. Now the work force of five generations have lost their identity but integrated with the local population identified as Assam tea garden workers/tribe equipped with infrastructure, health, and educational facilities for development and prosperity of the Nation, with committed, sincere dedicated efforts by the democratically elected popular, performing Governments of India.
5. These are the outlines and templates of the project envisaged and can be modified, improved upon, and fine-tuned with need-based adjustment on consultation with experts and administrative committees.
   1.1 The course of river Brahmaputra should be stabilized by periodic dredging, deepening, raising the bunds, and forestations with tall growing, sturdy timber trees up to 5–10 km in the boundary.
   1.2 The boundary of the riverbed should be strengthened by raising lemon grass, citronella, Pongamia, eucalyptus, Naraina (*Terminalia* sp) inclusive, at close spacing with deeper roots serving as a barrier against flooding the area, effecting the free-flow of water in the downstream.
   1.3 Horticultural crops like fruits and vegetables can be organized on a cooperative basis and precision farming. The harvested produce should be supplied to the warehouses and processing units and entrusted for marketing.
   1.4 River sand is very much in demand for real estates/realties. Supply can be initiated by ferrying the dredged sand to the

builders, earning good revenue for the exchequer and compensating expenses.

1.5 Many tea gardens of the tea industry, including the Assam Tea Corporation of India, are in crisis, due to various deficiencies, including financial crisis, maintaining the gardens with lack of depending work force settled in gardens for five generations. Restoration and revival of the plantations is a priority area under consideration of the implementation of innovative reforms, package of practices, mechanization, natural/precision farming, soil reclamation, consolidation, improved manufacturing, diversification, value addition, and marketing skill are envisaged for early revival and restoration of the plantations with minimum damage, adopting the technology and resources available in the country. The labour welfare measures ensuring their settlement through employment, creating avenues for job opportunities for the lakhs of able-bodied, trained, skilled, and dedicated work force on priority with justification. The required avenue and opportunities are to be created for enabling the millions for joining the mainstream is the top priority of the democratically elected government under the leadership of the Prime Minister of India. A few lines on the course of action, on priority with timeframes are narrated below for consideration, as a well-wisher of the prestigious tea industry and the nation.

1.6 The surplus workers can be settled in the bank of great river Brahmaputra valley of Assam with their consent where afforestation and horticultural projects are envisaged with processing and marketing facilities proposed to be created for the development of the North East.

1.7 Skill development training centers created for the willing able-bodied boys and girls to enable them to join working in the fields, warehouses, and processing factories including marketing through surface water transportation envisaged.

1.8 Cultivable land of about 2—3 ha per family could be allotted on lease for cultivating horticultural crops, fishing, and so forth for their livelihood. The organic produce can be procured by the cooperative body, ware housing, feeding the processing units established all over Assam, Majuli Island,

Gholahat, Tezpur, Guwahati and marketed availing the port and surface transport. The settled work force with land possession and liberty, can establish dairy farms, foliage cover crops, produce composts/cattle manures and supply to gardens, growers for promoting eco-friendly, organic good agricultural crops, which will be mutually beneficial for both the gardens and workforce earning their livelihood/ prosperity.

1.9 The North East has a fertile soil and climate suitable for various flora and fauna, and horticultural crops can be judiciously exploited, harnessing solar energy and environmentally friendly package of practices bringing more demand for cleaner value-added products globally for the welfare and prosperity of the nation. The new settlements created by the Government of India in coordination with the states of North East can be focused an example for the ecofriendly development of people below poverty line as a model village equipped with infrastructure, health, and educational facilities for development and prosperity of the nation, with committed, sincere dedicated efforts by the democratically elected popular and performing governments globally.

1.10 These are outlines and template of the project envisaged, can be modified, improved upon and fine-tuned with need-based adjustment on consultation with experts and administrative advisory committees.

4.7 Importance of maintenance foliage: intricacies on production and cost-effective management

The maintenance leaves in the canopy are the processing factories of the tea plant, producing the basic raw materials sugars and amino acids (assimilates), the template for the entire metabolism, productivity, and quality. Similarly the root system in the soil plays an equally important role of absorbing the nutrients and water regulating the assimilation using the natural resources like sun energy CO_2 from the atmosphere. The entire basic/ unique action of producing sugars and amino acids in the plant cell can be depicted below.

1. 6 CO_2+6 H_2O –hv/Zn, Cl-RuDPCarboxylase.–→ 3 PGAx2–→C6 H12 +6O2+8 ATP^
2. Alpha keto glutarate + NH_3———Transaminase—→Glutamic acid
   Glutamate + No3/NH_3–6-ATP/K2O/Mg–>Glutamine(Assimilates)*
   Asparate + NH_3—————Transaminase—————→ Asparagine.
3. Cysteine+HS/S/+NH3+ATPà-Cystine

4.7.1 North India

The principal parameters for the sustainable productivity of tea plants are indicated. Once the soil, plant, and ambient environment are optimized and balanced, the production is increased with quality. Cost of production comes down; value realization improves for cleaner produce; stress-free environment creates a happy win–win situation for all the stakeholders of tea plantations globally.

Nevertheless, the desired bush architecture is achieved/obtained with horticultural skill handling the plants from planting to formative prune. The process of building up the bush architecture has evolved with skill of workers and dedicated management of pioneers over the years. Judicious centering, tipping, pruning, and harvesting are involved. A minimum of 3–5 years is required for the formative prune. Being a perennial dicot, the longevity of tea bushes beyond 100 years assured, provided they are maintained free from undue biotic, a biotic stresses. Currently staggered pruning with rationalized inputs INM, and IPM on GAP is enforced for ecofriendly sustainable productivity with quality. Soil and plant biospheres, especially the canopy, health, and thickness, leaf addition, phasic growth, source-sink relationship, coordinating with dynamic physiological metabolic processes with need-based inputs on elite cultivars, ensuring timely effective harvest of the shoots, innovative processing, and online marketing with diversification and value addition hold the key for the success of tea plantations in the coming years. The prospective ambitious small growers of tea are likely to be benefited by the diversification and value-addition envisaged in the near future (Figure 4.3).

Strengthening and reclaiming the soil biospheres such as, humus, acidity, microbial population, feeder-root proliferation, balanced aeration and soil moisture, availability of macro–micronutrients play a role on absorbing the nutrients to sustain the photosynthesis, assimilation related metabolic enzyme reactions balancing with health, productivity, and quality of the produce impacting on the economic viability. The strength and improved health of the plants impart tolerance to pernicious pests and diseases free from stresses, increasing the cost-benefit ratio of the inputs reflecting on the economics/profit of the stakeholders; a win–win situation dreamed of

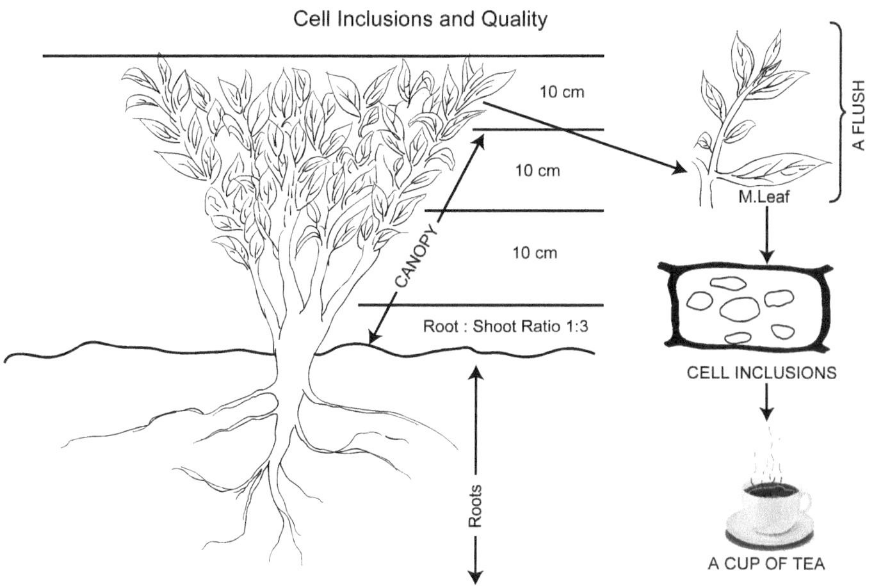

Figure 4.3 Impact of Bush Architecture on Cell Inclusions productivity and quality.

in plantation ventures. Thus, the key role played by the soil, plant biospheres, ambient environment through the regulated shade, cover crops, mulch, wind barriers with concomitant biodynamics of tea plantations in North Eastern India envisaged through implementation and enforcement of GAP, natural farming package of practices, concomitant with innovative reforms on administration, R&D, and welfare measures for the workforce by the Government of India will have salutary effects. Sooner the implementation of the restoration/reform measures and enforced, creating a congenial, friendly atmosphere, and cordial relationship between the management and the workforce with support, guidance, and monitoring of the progress, which is expected to reward the industry and workforce-depending population as well as the nation at large.

4.8 Shade management in tea plantations

Just like the soil/plant interrelationship, interdependence of shade trees and tea plant performance closely related with sustainability, productivity and interdependent, correlating, coordinating the dynamic metabolic activities. Thus shade management is almost the plantation management contributing the ultimate health, productivity, quality, and life span/

longevity of tea plantations. A three-tier vegetation in plantations: shade, both permanent and temporary; shrubs as cover crops, between hedges; herbs/grasses under the canopy since soft mulch is preferred, moderating and regulating the congenial ecosystem; and ambient environmental balance for the efficient performance of the tea plant, tying up the root/top metabolic activities ultimately determining the health and productivity of tea bushes.

The roots of shade trees help the pumping of nutrients from deep layers of soil, feeding the top canopy along with water and facilitating the photosynthetic capacity of maintenance leaves. The soil moisture and aeration in soil media with humus accelerate the microbial population and enzyme activities. The synergistic, symbiotic relation of mycorrhiza and root exudation accelerate/boost the photosynthesis and assimilation directly. It is a positive impact on the bud break, intense shoot growth/flushing, making the best use of the favorable ambient conditions created, reflecting on the health, productivity with quality.

It is established that more shade is better than no shade, especially under the presently changed climatic situations worldwide. An ideal shade should drop the leaves during winter exposing the bushes to direct light eradicating the lichens algal growth on bark, facilitating early bud break after pruning besides being legumes. The refoliation time should coincide with the bud break and flushing of the bushes, neither too early nor too late. The shade canopy is regulated before the monsoon by pollarding. The pollarding also contributes to the OM build-up besides the fuel requirement of workers.

The well-established shade trees of mixed species not only regulate the shade, ambient humidity, and leaf temperature maximizing the accelerated photosynthesis, but also pump out excess water in the root zone due to prolonged stagnation and water logging through transpiration. Provision of wind barriers on the windward direction, establishing fast-growing, fruit-bearing hardy trees, bamboo, providing shelter to fauna and flora, and maintaining the ecobalance, while cleaning up the worms and caterpillars economically. In addition the fuel baris, fodder grass in marginal fallow lands, and planting biomass producers will come in handy for compost and vermiculture with coexisting dairy farming for value addition.

Thus, growing shade trees for temporary as well as permanent use, and preserving the soft weeds not only moderates the environment conducive for growth and productivity, building up the humus content of soil, multiplication of microbial population; accelerating the root and enzyme activities. In a nutshell, shade management is plantation management, contributing the health and sustainable productivity, enhancing the economic viability of plantations; enabling the enforcement of ecofriendly GAP not only in Assam but also in hill ranges up to 1000 m elevation with advantage. Judicious diversification of intercropping with fruits and spices and value addition will further pave the way for prosperity and employment.

Agro-forestry, tea/coffee, rubber plantations, spice, and horticultural crops are complimentary to each other in nature for survival and mutual benefits, safe-guarding the ecobalance, environment, and biodiversity, contributing to the welfare of humanity globally. We have to live with nature and allow nature to flourish and nourish the depending population, adopting GAP/natural farming and avoiding all poisonous hard chemicals, in addition to preparing for natural calamities like flood, pollution, curtailing the industrial transformation of plantations, conserving soil and water resources (rivers), hills, and plateaus in the interest of future generations. We are responsible and accountable for any decision or action taken in this regard.

The time has come for selection of elite shade plants suitable for tea plantations of hills valleys, plateaus in the changed climatic conditions all over the world. This is an R&D priority to select/evolve and tide over the adversaries being faced by tea plantations globally. A combination of mixed shade with proper spacing meeting the future mechanization in plantations, suiting the region specific variables are to be borne in mind in breeding, selection, and release of any elite clone/cultivar on priority.

Some criteria for ideal shade tree for tea plantations are narrated for guidance, which depends on many factors such as judicious, need-based, situation-specific discretion for selection, evaluation, and release are envisaged. It will be helpful for most of the horticultural crops, though time consuming, it is possible with technology and resources available. The pioneers have done a wonderful job, and we have to carry forward with efforts on upgradation, fine tuning the breeding and selection of ideal shade patterns, and species and cultivars suitable at least for major regions of Asia (India, Sri Lanka, and China) Central Africa, Brazil, Argentina (Southern America), which are the major contributors of tea worldwide.

Parameters for an ideal shade tree for tea plantations:

1. Leguminous species preferred.
2. Mixed shades than single species desirable.
3. More shade is better than no shade can be regulated.
4. Drought prone hill ranges are benefited with tall growing species like *Albizzia lebbeck* at wider spacing, with shrubs and grasses, help against soil erosion and run off losses.

5. The periphery and borders should be planted with sturdy fast growing trees, and bamboo as wind barriers, while the interrows can be planted with grasses in contour-staggered trenches for soil conservation in steep slopes, moderating the ambient temperature, humidity, building up humus, and microbial population.
6. A three-tier vegetation with shade trees, interhedge cover crops, preserving the herbal soft weeds as mulch, helps against the soil wash/leaching of nutrients applied, conserving the soil moisture, build up of microbes, and proliferation of feeder roots, enhancing nutrient absorption and biomass production.
7. Depending on the seasonal changes and rainfall distribution, the shade trees should drop the leaves during winter and flushes back at bud break of tea bushes after pruning.
8. Intercropping with horticultural crops for diversification is beneficial and economical, cost saving.
9. Deep-rooted, stress-tolerant species/cultivar are preferred, for anchorage and effective exploitation of nutrients and water from deeper layers of soil. Many tea gardens/plantations, plateaus/river valleys in North India, and high range hills are exposed to floods, water logging during monsoons, and drought during winter. A legume tree with ideal effective good canopy and deeper root system will help these plantations.
10. Some proven combinations of shade trees, of three-tier vegetation are mentioned for guidance/trials.
    a. *Albizzia chinensis, A. lebbeck, Glyricidia maculata*: Assam, Arunachal, Plateaus and River valleys.
    b. *Albizzia odoratissima, A. lebbeck, Derris robusta/Leaucaena glauca*: Doars, Terai, Himachal, and Palampur.
    c. *Albizzia chinensis, Acacia lenticularis, Glyricidia, Erythrina, Grevillea robusta*: Darjeeling and the Nilgiris.
    d. *Albizzia lebbeck* selection, *Acacia lenticularis*, Grevillea, Glyricidia, Sesbania, and Tephrosia: South India and Sri Lanka.
    e. *Albizzia lebbeck* selection, *A. odoratissima/chinensis*, Grevillea, Glyricidia, Indigofera, Erythrina, soo babul, and bamboos: Nigeria, Central Africa, Argentina, Brazil, and Southern America.

Thus the shade, cover crops, herbs/grasses mulch, wind barriers, fuel baris, biomass producers, and horticultural crops play an important role in plantations, GAP/natural farming. This is a priority area of R&D to select the cultivar, evaluate their impacts/response in fields, develop packages of practices like population, spacing, and so forth to suit the emerging mechanization envisaged globally in the interest of producers and consumers, although the pioneers have done a wonderful job on ecophysiology of shade trees in plantations (Hadfield 1968; Margaret 1974) in India and Australia.

Given in this chapter are descriptions of the parameters of *A. lebbeck* tree growing wild in Coimbatore to decide on suitability as a shade tree with wider spacing, intercropped with fruits, spices, grasses, and herbs in mid—high elevations tea plantations with advantage of diversification and value addition, adopting ecofriendly GAP/natural farming and mechanization.

Parameters: A. lebbeck. Select: 20 years old, self-maintained trained, 10 m tall—7 m trunk, and 3 m canopy branched ideally 1:3:5 with dark green foliage. (Fig) Stress tolerance, suitable for hill plantations.

This is a stress-tolerant, leguminous, early sleep and rise, harboring birds, suitable for wider spacing complementary with fruits, spices, herbal crops with wind barriers, hedges in periphery, grasses like Vettiver, citronella, and lemongrass as binders against soil wash, leaching; conserving the soil, nutrients, and enhancing the humus content and buildup of microbial population. It seems a potential shade tree for hill stations in combination with other horticultural crops and biomass producers for ecofriendly/natural farming with advantage. Worthy of introduction and field trials taken up to evaluate the impacts and developing package of practices before release.

A Popular shade tree with optimum, space, canopy by pollarding in pruned tea. *A. odoratissima, Leguminaceae.*

A pruned field in Gantapara, TE. Doars, West Bengal of Gooricke Tea Ltd., with a good stand of *Albizzia odoratissima* shade, with optimum spacing and population of both tea and shade while saving the prunings.

Typical shade for South Indian tea plantations: *Grevillea robusta.* A three-tier shade pattern with *A. mollucana, Glyricidia,* and *Dadaps* (Erythrina) used with periodic pollarding with advantage in SriLanka.

Albizzia lebbeck, new selection for high elevations, stress tolerance for evaluation and decision.

Fig: *Albizzia lebbeck* selection 1/Tephrosia Purpurea herb to be inserted.

Fig. above: A kolinchi (*Tephrsea purpurea*) annual, ideal for rehabilitation and cover crops.

Fg. below: Grasses Vettiver and citronella for rehabilitation.

Vettiver, Citronella grasses, and Kolinchi (*Tephrosia* sp) are good for soil-binding, enrichment, and reclamation.

The photo depicting the parameters required for young tea in Assam. (Albizzia- Odoratissima shade, drainage, and a growing weather and soil in Assam valley), Thowra TE, Warren Tea, Ltd. Consolidation of shade trees, temporary shade like Indigo fera, Goraneem, and deepening of peripheral drains required to improve the microclimate, reducing the inundation/water logging during monsoon, for sustaining the growth and productivity taking advantage of the peak season. Clayey heavy soil, water logging, consequent root diseases are the principal problems in both young and mature tea besides the brick kilns nearby. Prone for severe red spider mite incidence during winter, aggravated by the dust from the highways, brick kilns stagnating the crop level around 2000 kg/ha despite the ideal upper Assam growing weather (Manivel visit report, 2009—10). Marginally high pH, water logging/Inundation due to depressions, incidence of root diseases, corticium, and red rust were tackled with ground application of dolomite 100 kg mixed with biocides trichoderma, pseudomonas, and paecilomyces at 2 kg each/ha and applied in the soil and incorporated with hoeing in the month of December when soil was moist. It was found that the bushes recover by March, indicating the maladies were taken over by the soil amendment, biocides, oil cakes like neem, castor, mustard, and pongamia accompanied by hoeing which induced the root aeration promoting growth. It was clear that the package can be adopted all over Assam plantations with the advantage for GAP/natural farming.

Albizzia lebbeck selection 1 is an ideal alternative shade for hill plantations of North and Southern India.

Citronella, lemon grass, and *Vettiver zizanoids* are very good binding grasses in the borders of sections and stream boundaries, minimizing soil erosion and run off/soil wash. These grasses include *Tephrosia purpurea* and legume annuals that can be used for quick reclamation of soil.

4.9 Stress management

Tea Plantations all over the world go through many stresses depending on the soil, terrain, rainfall distribution, geographical location, cultivar, field management practices, population density, and inputs. These are narrated with causes, and suggested ameliorative/palliative measures.

These can be categorized as biotic and abiotic.

1. Biotic: Pests and diseases
    a. Pernicious pests: Red Spider Mites, Helopeltis (Tea Mosquito Bug), Semiloopers (*Hyposidra talaca*), red slug, mealybug, and scales, termites, nematodes, etc. Virulent resurgents: jassids, thrips, and aphids.

b. Diseases: Frame and leaves/canopy: Blister blight (*Exobasidium vexans*), Black Rot (*Corticium theae*), Red Rust (*Cephaleurus Parasiticus*), Gray Blight (*Pestalotia* sp), and frame canker (Poria, Pomopsis).
Root diseases: Brown Root rot (*Ustulina* sp), and fusarium.

2. Abiotc: Water logging, drought, hail storm, frost damage, toxicity, residues, chemical pollutants

3. Other extraneous adversaries and natural calamities

4.10 Biotic stresses

The ideal conditions of soil, plant, environment, soil moisture, aeration, and inputs play a major role on the health and productivity of plants. Any deviation from the norms will have impacts on the bush activity growth and development depending on the severity of deficiency or excesses. The plants sense the adversaries and react showing distress symptoms. Thus weakened bushes become vulnerable and predisposed to pests and diseases affecting their performance the most serious in Assam tea is the recurring floods, Extended water stagnation in the fields during monsoon downpour. The root absorption of nutrients with water hampered affecting the photosynthesis and assimilation of maintenance leaves. The root starch reserve is depleted and weakened. These weakened bushes become vulnerable to red spider mites (RSM), which prefer and thrives on dry leathery leaves, sucking the sap for nourishment and turning the leaves brownish. Similarly, lack of aeration burns up the starch reserve and weaken the bushes predisposing to root rot diseases and thus the affected root system will not be able to absorb the nutrients along with water, and support the metabolism of the plant canopy. Thus, the incidence of pests and diseases are aggravated by the ineffective drainage, prolonged, water stagnation, and recurring floods in Brahmaputra River every year changing the course.

Once the course of the Brahmaputra River is stabilized and flooding prevented by the joint ventures of plantations, local authorities and central government, there will be less adversaries with negligible incidence of maladies and requirement of remedial measures. The impact of maladies will not be severe and can be managed with available technology "integrated nutrition management, integrated pest management and good agricultural practices". Foliar application: implications and salient guidelines for tea plantations.

Nutrient foliar application/feedings is not a substitute for ground application. However, it is a potent tool for need-based quick recovery of plants from various stresses and amelioration measures. Foliar absorption takes place through stomata distributed on the under surface of leaves. Stomatal opening and closure regulated by light intensity and K level in guard cells. In addition to stomata, absorption can take place through cuticle cracks, lenticels in the stem bark etc. soil moisture and ambient humidity also play a role in absorption rate of any foliar. Time and dose play an important crucial role on efficiency. Morning and evening with enough sunshine are found favorable for absorption. Mid-day closure of stomata should be borne in mind in any foliar application. In addition to the individual valence of the elements, their interaction-synergism and antagonism play a moderating role on absorption. It should be need-based rather than luxury. Continuous use of any foliar impacts on addiction kind of inbuilt rhythm. As a result the effectiveness gets reduced. The zinc plus triacontanol foliar for growth promotion(synergism) under light limited field situations rectify the addiction and ameliorate. The isotonic effect of absorption of NK foliar at 1% was observed in the field situations. Similarly the same has been found in zinc absorption in lab studies. For any foliar a minimum gap of two hours of rain-free sunlight is required for 50% completion/absorption. In case of rain within two hours, the spray should be ignored or repeated.

The common, popular foliar schedules of tea plantations management are listed below for information and guidance for Assam, North East plantations. (Vide Appendices, The calendar of foliar operations Recommended for Halmari, Mokalbari, Khongea, Boisahabi, Bukial TE Assam, and North East tea gardens.)

1. NK foliar for drought/stress amelioration at 1% each during winter overcoming the moisture stress.
2. Micronized micronutrients, Zn, Mg, Mn, B, and S with organic growth promoters/biostimulants to promote bud break, growth, and sustained productivity with quality, including wound healing after hail damage.
3. Zinc plus triacontanol to increase the photosynthesis of maintenance leaves compensating the light deficiency due to overcast cloudy weather during peak monsoon seasons.
4. Combinations of biocides(bioconsortiums) for managing the pernicious pests and diseases with GAP, strengthening the soil, plant biospheres ensuring sustained higher productivity.
5. Fertigation in young tea, nurseries, intercrops planted for diversification/ value addition, etc.

Fortnightly foliar of consortiums of micronized (nano)micronutrients with proven biocides on machine-harvested quality gardens/fields for sustained health and productivity, preventing the incidence of gray blight.

CHAPTER FIVE

Pruning systems and crop productivity

Contents

5.1 Normal pruning, rejuvenation pruning

5.1.1 Concepts

The three processes such as nutrient absorption by roots, photosynthesis by the maintenance leaves, and assimilation are energy dependent reactions that are very strongly related to the productivity of the tea bushes. Pruning strives to maintain a balance and coordinate between these metabolic processes for sustainable productivity with cost economics.

5.1.2 Normal pruning

Periodic pruning restores the rhythm of flushing, breaking the apical dominance and inducing the dormant buds from below to grow, which restores the vigor and size of shoots. Pruning involve judicious knifing once the plants have been established in the field. Training is done in the formative stage within three to four years of planting, while pruning is done periodically, once every three to five years, to maintain the harvesting table (bush height). Cutting the top hamper also regulate the number of primary stems and buds for cropping, restoring a balance between the bush capacity/potential and the sustainable productivity. Illustrations on training and pruning (sketches) are shown later in the chapter (Figure 5.1).

Scientific Perspectives of Tea Plant Horticulture and Productivity
ISBN: 978-0-12-823444-0
https://doi.org/10.1016/B978-0-12-823444-0.00005-4

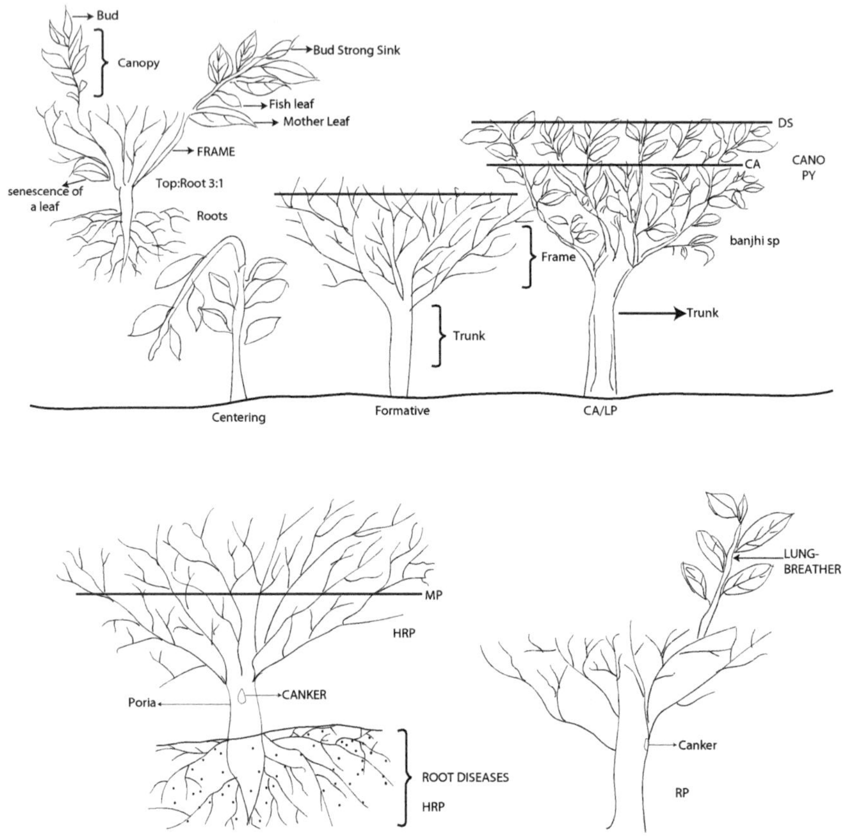

Figure 5.1 Bush architecture- intricacies of pruning.

5.1.3 Other knifing processes

Historically bushes had been pruned in summer (June) annually every year up to 1964. This led to bark scorching, frame canker, and yield reduction. The harvest index was around 7.5%which is considered low (Barua, 1989) The pioneers therefore shifted to staggered pruning-LP/CA-DS-UP-LS-UP or LP-UP-Ds-UP for the crop and quality with harvest index up to 11% and sustained with additional inputs. In course of time due to overexploitation with intensive cultivation measures the bushes exhausted, showing fatigue. The weakness with deficiencies in soil and plants led to dieback and vacancies. These setbacks of plant health and productivity combined with quality deterioration and value realization forced the plantations to go for rejuvenation pruning.

Starch reserve in roots and frame plays an important role on the bud-break and recovery of the bushes. A minimum of 4–7 weeks resting of the bushes is recommended to ensure the buildup of the starch level. The period of resting depends on the type of pruning/severity (depth) of cut and the minimum level of starch reserve is 15%–20%. The threshold level of starch for surviving the metabolic activity without growth falls around 12% which varies with terrain and climate of the regions.

5.1.4 Rejuvenation pruning

Tea bushes are perennial dicot with vascular bundles of xylem and phloem mediated in between the cambium, the most versatile tissue contributing for annual growth sustaining the growth and perpetuation. Thus logically all the perennial trees have no age, blessed with the capacity to perpetuate indefinitely. Nevertheless the growth and productivity of tea bushes are governed by many factors with intensive cultural and management practices, going through many stresses of varying degrees, affecting the growth and productivity, and thereby the lifespan, of the bushes.

There are gardens even 100 years old giving satisfactory crop with quality in different tea regions. However, from 1964, lot of refinement/modifications in bush management including staggered pruning, increased bush population, intensive cultural field management practices; increasing the dose of chemical fertilizers and plant protection chemicals, weakened the bushes, and predisposing to various stresses, making the operations uneconomical and viable.

After a span of about 50 years of intensive cultivation of tea the soil health, environment, and water sources have been deteriorated, contaminated, and polluted beyond redemption, culminating in stagnation of productivity, poor in quality, and value realization. Two options are left to the management under the situations—either to abandon or uproot and replant. To overcome the situation a midway option like rejuvenation and consolidation is considered for reviving the health and productivity of the bushes.

5.1.5 Criteria for rejuvenation

Age above 50 years, vacancies more than 30%, productivity below district average affecting the economic viability, evidence of root diseases, frame canker, and degraded unhealthy soil resulting in high vacancies of both tea and shade trees will qualify for rejuvenation pruning. Fields with unsuitable cultivars, such as extreme China, and susceptible new introductions are also considered for rejuvenation and renovation restoring the productivity and economic viability of the plantations.

5.1.6 Steps/salient points on rejuvenation

Uproot and deep plow to breaking the hard pan, exposing the soil through solarization. Correct the soil pH through soil amendments and optimize the organic matter (OM) level of the soil. Rehabilitate with Guatemala grass/ Mimosops planting, slash, and incorporate for OM.

Normal recommended period is three years with Guatemala or Mimosops. But the gestation period for rejuvenation three years and new planting five years (eight years) became uneconomical. Hence, upgradation of rehabilitation/revival process deserve to be reduced to one year, avoiding deep till aging by substituting potent biofertilizers/consortiums besides using prospective annual/biennial herbs like *Tephrosia purpurea* for rehabilitation are being explored (personal communication).

Level and mark the positions of bushes and shade trees as per layout.

Plant and establish the permanent shade trees.

Ensure filling up of required inputs for successful establishment of both shade and tea plants.

Complete planting of tea (double hedge, staggered; around 15,000/ha) avoiding monsoon.

Establish temporary shade as well as leguminous cover crops for moderation of temperature, Mulch the soil avoiding the collar region for controlling soil water evaporation.

A Pruned field with pruning conservation with good stand of shade in Assam/Doars

5.2 Bush architecture: pruning, tipping, and harvesting

Bush architecture plays an important role deciding the increase in crop besides the cultivar and population/density. Low frame/volume with adequate number of sticks primary and secondary minimizing the distance from root, to growing shoots through the frame and canopy is recommended, sustaining the flow of assimilates to the growing shoots with the bud, which is the strongest sink. The canopy health and density of the foliage determines the density of shoots and growth rate. The top 10 cm of the canopy contributes about 50% of the production; the middle layer of 10cm of canopy contributes 25% of the productivity of shoots. The bottom layer of 10 cm serve as moderator/buffer against various stresses and marginally contribute especially during the favorable peak cropping season. The photosynthetic efficiency of mature tea leaves attains the maximum (12−15mg/dm2/h) within a month of unfolding and continues up to six months. In cooler regions like Darjeeling the chinary cultivars retain the leaves for longer periods, though there is reduced pn, due to age, self-shading, and weakness. For maintaining the health and productivity, many measures like leaf addition, pruning time and measure, periodic harvesting, and foliar feeding are encouraged judiciously to sustain the efficiency and health. Once the leaves become less efficient, stagnation of production, need-based steps are taken for leaf addition either mother leaf or fish leaf for augmentation; bearing in mind the creep and quality. The older lower layer of leaves contribute mainly for enriching the starch reserve in roots and senesce after a year or so of emergence. The senescing leaves translocate the constituents to the needy neighbor and drop off. A mature maintenance mother leaf once attains maturity exporting the photosynthates to other needy parts, like buds, shoots, flowers, fruits, and roots will never import even when starved. Instead they senesce and safeguard the bush from any stress.

The other steps of augmentation, restoring the health of the maintenance canopy is staggered pruning of 3−4 years cycle, balancing the crop, quality ensuring the productivity; crop distribution, inputs, cost of production (COP), etc. The recommended pruning cycle depends on cultivar, elevation, terrain, and climate of the region concerned. The popular staggered pruning cycle in North India are given in this section.

LP/CA-UP-DS-UP/LP/CA-DS-LS-UP. Four year cycle.

The quality conscious planters opt for a three-year cycle, especially for China hybrids. The staggered pruning cycle maintaining a healthy canopy besides streamlining the manurial inputs, rationalizing the dose, splits, mechanization of harvest during peak season, supplementing with foliar application of micronized nutrients, getting favorable crop distribution, integrated with INM and IPM for cost-effective management results in sustainable productivity with quality. Although this package has been accepted in principle, further fine-tuning, refinement on diversification, and value addition are paving the way for good agricultural practices (GAP) with advantages for production and direct marketing of specialty teas by the emerging small growers is a sign of progress for the Indian tea.

South India, Darjeeling, and Sri Lanka tea.

These three regions of tea follow longer pruning cycle of four to five years of LP/CA, suiting to the lateritic ferrous soil rich in organic carbon-humus (peat soil) undulating terrain/slopes, and distribution of rainfall thereby the crop as well. The pruning time coincides with the lull period in April and August. During the unpruned year, new mother leaves are added to the plucking table periodically and judiciously, filling up the gaps and cavities forming the new table of maintenance leaves to sustain the productivity and health of bushes. Thus the importance of maintenance leaves amply recognized which has a definite bearing on the response to the inputs, productivity, quality, and incidence of many pests and diseases and stresses.

The manurial doses of NPK recommended without any cap/ceiling and enforced in the 1970s and 1980s based on the replacement theory, which caused enough damage to the soil and bushes irrevocably, reducing the productivity, quality, and escalating the cost of production. Emergence of virulent pests like helopeltis, thrips, and red spider mites due to aggravated stresses, weak unhealthy soil, and bushes, which is consequently a critical adverse situation never experienced before, warranting many reformative measures, fine-tuning, rationalizing the field management package of practices for rectification and restoration of soil, plant health, and productivity.

1. Integrated nutrient management retaining the NK maximum to 300 kg per ha per year at 2:1:2 ratio with three to five splits balanced with adequate OM, restoring OM content and build up of microbial population, thereby restoring the health and productivity of the bushes. The strength and health of soil and plant (biospheres) contribute to sustainable productivity with quality besides the timely periodic harvest of the shoots, fixing the fines around 65% manufacturing without RC, will

pave the way for speedy recovery and prosper. Biospheres of root and foliage canopy are interrelated and coordinated for effective utilization of inputs for cost-effective management. Enforcement of good agricultural practices are the only way. Hyper acidity, low OM, hard pan, low microbial population, poor starch reserve in roots, emergence of virulent strains of pests, and diseases due to continued use of hard chemicals and herbicides are responsible for the present maladies of South Indian Sri Lankan plantations. Adoption of ecofriendly GAP is the only way to recover from the adversaries to prosperity with necessary required technology and inputs are available in the country for GAP, precision, and natural farming, especially for the small growers with diversification and value addition with advantage.

These enlisted adverse soil, plant variables reflect on the productivity, quality, crop distribution, price realization of North and South Indian tea. North India produces about 3000 kg/ha of supreme quality tea in flushing period of about eight to nine months, while the South India produces almost the same crop throughout the year, but poor quality due to coarse plucking and their manufacturing practice and style of recycling.

Thus, there is tremendous scope of improving the productivity, quality, fine-tuning their field, and processing practices on priority for enhancing the image and economy of South Indian tea plantations.

Consequent to recommendation and implementation of packages for GAP, optimizing the soil, plant ambient parameters (biospheres), and inputs, the plants are healthy and strong, less vulnerable to biotic and abiotic stresses, assuring sustainable productivity with quality and cost of production being affordable. This is a welcome sign of restoration from the damages the plantations gone through almost in the past 30 years due to indiscriminate heavy use of hard chemicals. The turnaround is around the corner, anticipating 50% crop increase in the coming years with good quality and value realization.

CHAPTER SIX

Physiology of the tea plant

Contents

6.1 Cultivar with varying harvest index

The cultivars in the fields play a definite role on the productivity, quality, and cost economics. At least in new plantings, proven clones and biclonal stocks should be encouraged and recommended for planting. TV 1, P 126, AV 2 (chinary); TV-17, T3E3, S-3 A-3, TV-20 (Assam) TV-9, TV-23, TV-25 (Cambod), TS 462, 491, 520, 581, TRF-4 (Biclonal Stocks).

6.1.1 Cultivar and rootstock

It has been found that the leaves/canopy controls the mechanism rather than the roots. This had been demonstrated in Tocklai, by top working of Assam (TV20), Cambod (TV18), and China hybrid (TV1) cultivars over a seed jat

Scientific Perspectives of Tea Plant Horticulture and Productivity
ISBN: 978-0-12-823444-0
https://doi.org/10.1016/B978-0-12-823444-0.00006-6

(70 years old plant). In a common environment soil, root stock, the cultivars exhibited their in-built behavior of bud breaking early or late after winter dormancy independently and continued the rhythm until the end of December. This had been confirmed by reciprocal grafting of the clones. The genetic makeup (DNA), mediated through the proteinaceous phytochrome/cytochrome P 680−700 in plant cells, chloroplast plays a vital role regulating the internal medium and constituents (The Hindu Science & Technology, November 21, 2019, quoting Nature, 2019).

6.2 Carbon metabolism: photosynthesis and assimilation

Crop productivity depends on photosynthesis of the maintenance foliage and partitioning of assimilates in to growing shoots.

Photosynthesis and assimilation: The photosynthesis and assimilation thereby the productivity and quality are mainly governed by the light intensity, temperature, and humidity of the ambient area around the canopy in the fields. This is further augmented by the nutrient supply, mulching, and soil moisture. The health of the "maintenance" leaves, active growth, and spread of feeder roots coordinate the assimilation of photosynthates to sustain the growth and productivity of the bushes.

The source-sink relationship, flushing behavior, the banjhi proportion, cropping behavior, and distribution, apical dominance, phasic growth are governed by the hormonal balance between the promoters and the inhibitors, which are influenced by the environment and health of bushes including the cultivars' genetic makeup, both interrelated and correlated.

6.3 Source-sink relationship

Source-sink relationship in a tea plant plays an important role, ultimately determining the productivity with quality and sustainability.

In simple terms, the source is the producer, manufacturer, supplier while the sink is the user. In a tea plant the producer of assimilates (sugars and amino acids), consisting of mature maintenance leaves (canopy) is the source/supplier of food.

The root, trunk, and frame are the users, the sink. The roots are the strong sinks accumulating the starch reserve supporting the metabolic functions even in winter when photosynthesis (sugar synthesis) is minimal for subsistence, as well as when the bushes are pruned. The developing buds of tea shoots serve as strong sink in early bud stage, draining the photosynthates heavily for their development. As the bud opens up and grows into shoot, 2+bud, the sink capacity declines at 30% for each leaf gradually. Thus the sink capacity of the buds and the source capacity (photosynthetic rate) of maintenance foliage are governed and mediated by the process of harvesting flushing succulent shoots, impacting on the productivity and quality. The tea bush is very often considered as sink limited rather than source limited. Thus judicious, periodical, timely effective harvesting of shoots plays an important role on sustainable productivity and quality, besides other inputs.

In a pruned and plucked tea bush, the source is fixed while the sink capacity is varying due to periodic harvest of shoots. The regeneration of buds also takes some time. As the new leaves unfold from the bud, the bud loses the sink capacity, and in transition it becomes a weak sink and self-supporting. The tender succulent shoots (2 + bud/3 + bud) are harvested for making the tea of commerce.

Thus, the tea bush is sink limited rather than source limited. The sink-induced photosynthetic efficiency can be sustained and balanced by judicious plucking of shoots retaining the buds, 1 + buds, harvesting through standard plucking of 2 + bud or 3 + bud. Black plucking, which is removing everything above the table, affects the photosynthetic efficiency of the maintenance canopy and underutilized, losing the crop even under favorable situations.

Depending on the relative strength, the pulling force and direction of movement of metabolites from the maintenance leaves (canopy) is determined. The quantum of flow of metabolites (sap) determines the number of shoots grow. Thus, the phasic growth, flushing behavior, and productivity are governed by the source-sink relationship and hormones.

6.3.1 Relationship between source-sink and hormones

The source-sink relation is based on the Malthusian theory of demand and supply, regulated. The source-sink relationship is governed by hormonal balance and the demand and supply within the bush. Auxin in the shoot tips, gibberellins in the maintenance leaves, and the cytokinins in the roots are the hormones involved in regulating the source-sink relationship mediating the flow of metabolites, leading to the intense cell division and multiplication of the shoot tip.

This process is augmented by the synergistic, correlation effect of cytokinin, synthesized and migrated from the roots. Once the new leaves unfold from the bud, the sink capacity of the bud gradually declines, culminating into a banjhi shoot unless harvested. Thus the process of plucking ensures the regeneration of shoots besides giving the valuable crop.

Very often the bushes are sink limited rather than source limited due to periodic plucking/harvest of shoots. Thus, we are sacrificing the 12%–15%, sink-induced photosynthesis due to periodic harvest of growing shoots for the production of the tea of commerce. Some ameliorative measures like foliar application of ecofriendly biogrowth promoters and nutrient supplements are adopted to minimize the impact.

Illustrations (graphic). (1) Response of intact maintenance leaves on photosynthesis rate by 1 light intensity, (2) temperature, (3) CO_2 concentration, (4) source-sink relation, and (5) harvesting.

6.4 Apical dominance

Apical dominance plays an important role on phasic growth of shoots and crop distribution. Apical dominance and the bud dormancy are in-built safeguards against the stresses, sustaining the productivity. All the buds in the bush never grow at a time, some proportion/percentage of buds will go banjhi (quiescent), maintaining the bush health and preventing from exhaustion and safety from the stresses.

Even during growing monsoonic weather, 10%–12% banjhi shoots are evident indicating the bushes go through certain amount of deficiencies or stresses, 100% luxuriant growth is not possible in nature.

The shoot apex dominates the lateral buds in growth; forcing them in to quiescence/dormancy in tea plant. This phenomenon is governed by the auxin and indole acetic acid (IAA); the concentration is its balance with growth retardant like abscisic acid/ethylene. High level of auxin in the tips triggers synthesis of ethylene which can act synergistically based on the concentration and weather.

6.4.1 Overcoming apical dominance

The apical dominance is broken by removing the tip, breaking the stem, pegging, debudding the main stem, or by using plant growth promoters like gibberellins. In weak plants and seedlings, induction of laterals is a bit difficult and erratic due to the genetic complex, where all these measures are used to induce lateral bud break, breaking the apical dominance.

Tipping the shoots periodically induces branching and spread of the frame, building up the top hamper/canopy. By raising the tipping height (35, 40, 45, and 55 cm) yearly, the plucking table is formed at 90 cm a convenient height for the women workers to harvest the crop. Formative pruning is done in the third or fourth year depending on the growth, when final cleaning, desnagging, and opening the center, fine-tuning the frame.

6.5 Flushing behavior and hormone relationship

6.5.1 Endogenous hormone

Auxin primarily of IAA and gibberellins that are light dependent and are produced in leaves and responsible for the cell multiplication and elongation. Cytokinins are mainly produced from roots and translocated to the shoot buds, regulating the cell multiplication and growth. Thus, these three promoting hormones act synergistically to regulate the bud break and growth. Their concentration as well as the balance between them play a role in response to the bush health and environmental factors. At times, especially on pruned bushes, multiple shoots are produced.

This may be due to the increased vigor and less competition. It is not clear which component of the controlling factors that are responsible for this kind of mechanism. It is a clear indication that tea bushes can be induced to produce multiple shoots thereby many fold increase in crop without compromising the sustainability and quality of the made tea.

6.5.2 Maintenance foliage and hormone

The maintenance leaves in the canopy are the sensors of the environment through the in-built phytochrome of the bushes and regulate the flushing behavior of the bushes. The maintenance mature leaves, regulate the bud dormancy through the proteolytic phytochrome in chloroplasts of the leaves. The phytochrome senses the environment that triggers the enzymes responsible to synthesize the hormones, auxin, gibberellins, abscisic acid, and activate the phasic growth of the bushes.

6.6 Winter-bud-dormancy-hormone

The bud dormancy and interflush dormancy/quiescence are in-built and can be modified or period reduced but cannot be completely eliminated. The bud dormancy/interflush dormancy has a positive role to play on the health and longevity besides safeguarding against various stresses. As a result of phasic growth habit, the bushes go through interflush dormancy-lull periods.

The main hormone responsible for bud dormancy is abscisic acid and its concentration. When the winter breaks with showers, weather warms up, in February/March, the leaves synthesize gibberellins also. Thus the balance between the abscisic acid and gibberellins (GA/ABA) determine the bud break commencing the new growth. The clones representing Assam, Cambod, and China differ in their response to low temperature and recovery. The Assam recovers first followed by Cambod and China setting their flushing behavior and crop distribution.

Tea bushes in Assam go through winter (December—February) shutting down the growth and allowing for important field operations like drain cleaning, pruning bushes, hoeing and weeding to loosen the surface soil, exposing the soil-borne worms, at times breaking the hardpan and infilling of shade and tea vacancies. Although the maintenance leaves are green, the metabolic functions are at low key for sustenance. The buds are dormant due to low temperatures (5—16°C) building up the starch reserve in roots.

6.6.1 Factors responsible for flushing behavior

The intensity and period are governed by the cultivar, moisture stress, soil and climatic factors, sap flow (vascular connection), nutrient deficiencies, starch reserve status of the bush, and ultimately the hormonal balance between the promoters (auxin, gibberellins, cytokinin) and inhibitors (abscisic acid, ethylene) in the leaves forcing the buds dormant/quiescent and the length of period.

Photo of a top-worked bush on flushing behavior.

1. Bhanjhiness versus the crop distribution in Assam.

6.7 Starch build up in relation to flushing behavior and in-built mechanism in the canopy for sustenance

During active growth and monsoonic weather/season, the bushes maintain the starch level around 15%—16%, for sustainable productivity depending on the locations (terrain, climate) as well as the cultivars.

Main causes for the poor starch reserve in the root systems are poor frame, canopy thickness, inadequate maintenance leaves, diseases, stresses due to excess water—water logging or deficit water, drought, inadequate nutrition, and overexploitation.

If root starch level by any exigencies reaches below 10%—12%, the growth stops, affecting the metabolic level to the minimum for sustenance.

If the situation is prolonged, the upper leaves of the canopy support the top hamper while the bottom/lower leaves support the root to maintain the sustenance level 10%—12%. If the stress prolongs further the lowest leaf in the canopy senescence, translocating the metabolites in the leaf to the upper top hamper for survival. The DNA and phytochrome sense the gravity of the situation and activate enzymes responsible for abscisic acid production, which in turn effects senescence of the old exhausted leaf abscise and drop safeguarding the life of the plant (Figure 6.1).

***Sketch showing the direction of movement of photosynthates from the maintenance leaves. (Canopy** and Crop Distribution).

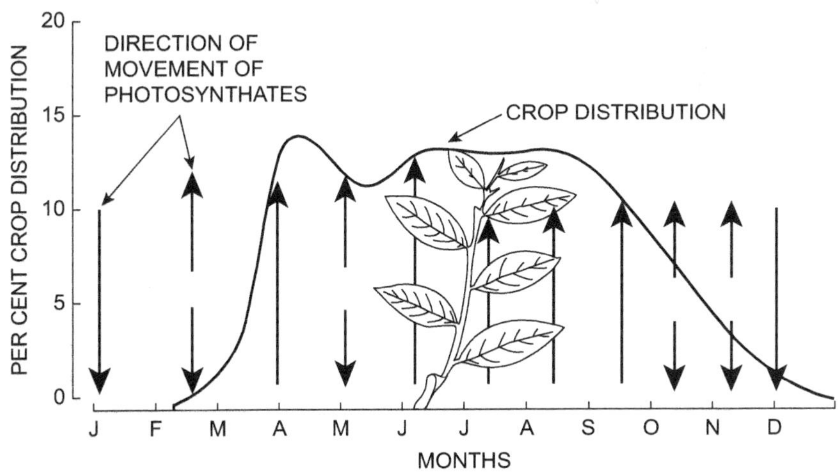

Figure 6.1 Movement of photosynthesis and crop distribution.

6.8 Wind, hail, and flood in relation to physiology

Speedy winds impact the stomatal closure, interfering with the gas exchange. Both transpiration as well as photosynthesis are affected impeding the productivity. A reasonably strong wind barrier of about 100 m width in the periphery preferably on the windward direction will be helpful in reducing the wind speed.

Occurrence of hail in plantations is common (April to June) before monsoon, causing widespread damage in plantations of North India. Uprooting of shallow rooted shade trees, shredding, defoliation of maintenance leaves of the canopy, and peeling of bark happen. The wounds due to injuries should be protected from pathogen entry. Simultaneously, foliar feeding of the bush with PGR, biostimulants, and cocktail of biocides like *Trichoderma* and *Bacillus subtilis* within three days of occurrence will minimize the damage promoting speedy recovery. This is repeated after a fortnight.

Flooding of plantations occur during monsoon season (June–September) inundating the plantations. Although the water is drained through streams and seepage, the roots are invariably submerged. Prolonged submersion affect the aeration of the roots hindering the absorption of nutrients and enzyme activity.

Starch reserve in roots will be utilized and exhausted for the support, weakening the bushes and making them vulnerable to root diseases. It had been observed that inundation for 24 h hurt the plants, exhausting the reserve; when inundation continued for three days the bushes have completely exhausted the reserve, inflicting irreversible damage and collapse, as revealed by the dynamic status of shoot water potential in lab studies.

6.9 Stress management in plantations

Tea, being a mono crop, is raised depending on the distribution of rainfall and the total rainfall, and the plantations are exposed to various biotic as well as abiotic stresses.

Both excess and deficit soil moisture affect the metabolic functions of the plants, predisposing them to stresses affecting the productivity and quality. Intensive cultural practices adopted in the field management of the crop also affects the soil media and environment making them vulnerable to pests, diseases, toxicity and or deficiency of mineral nutrients. Also, as a consequence of intensive cultivation, adopting excessive use of chemical fertilizers, the soil becomes acidic, exhausting the organic matter (humus) in the soil.

Soil moisture and ambient moisture content (humidity) reflect on the shoot water potential, regulating the stomatal pore opening. The gas exchange, through transpiration and photosynthesis, maintains the leaf temperature, governing the photosynthetic rate of the leaves in the canopy and thereby the productivity.

Desirable leaf water potential and the stomatal opening can be obtained through subsoil irrigation, drip irrigation, sprinkler irrigation, and mist blowers to tide over the moisture stress and keep the stomates open. Thereby, the gas exchange is moderating the leaf temperature, sustaining the Pn and productivity.

6.10 Secondary metabolites

The main stresses the tea plantations go through in Northern India include the following.

Drought (moisture stress) in winter forcing the bushes to go quiescent, and bud dormancy (banjhi) helping for pruning of the bushes. The desiccant foliage of the bushes, and dusty environment, attract the red spider mites.

Water logging (excess moisture) during monsoonic summer is conducive for the root rots and foliage (black rot) diseases.

The bright sunshine accompanied by high leaf temperatures beyond 35°C impede the photosynthesis of leaves, warranting shade regulation, since both dense and low shade are harmful.

Erratic crop distribution with peaks and troughs, as well as undue pressure on the management, work force, and factory, for effective harvest and processing of crop produced. Gardens in hills like South India, Sri Lanka, and Darjeeling face the problem of blister blight (*Exobasidium vexans*) due to prevailing high humidity.

Occurrence of adverse soil pH due to floods and excessive use of fertilizers/chemicals/herbicides affects the efficiency of nutrient absorption, exhibiting nutrient deficiencies like K, Mg, Zn, Mn, S, and B affecting the productivity, warranting for soil amendments during winter.

Intense flowering and fruiting draining the energy indicating the stresses in young tea and weak bushes.

Defoliation of mature bushes expressing distress due to stresses, toxicity, and deficiencies of higher vacancies beyond 30%, of both tea as well as shade.

Severe, acute Zn, Mg, Mn, S, and B deficiencies leading to defoliation dieback and vacancies is a common sight in many plantations of Assam, Doars due to indiscriminate excessive use of hard chemicals especially the glyphosate in recent years.

Stresses in the plantations vary in type and intensity depending on the geographical location, climate, soil, and environment, and the type of management practices adopted, inclusive of the inputs, quantum applied, etc. Thus prevention of any stress through prophylactic measures is better than palliative measures saving the valuable resources and the health of the bushes.

6.11 Remedial/restoration/palliative measures contemplated as scientific, financial, and sociological

Remedial/restoration/palliative measures contemplated both scientific, financial, and sociological measures. The principal restoration measures on priority are listed below:

1.
   a. Optimize the soil conditions: pH, organic matter, and microbial population.
   b. Optimize the bush population and shade by consolidation/infilling.
   c. Rationalize the soil moisture and aeration (60:40) promoting the feeder root proliferation and nutrition absorption.
2. Retain/preserve herbal soft weeds as a mulch, saving soil moisture, and build up beneficial microbes.
3. Prevent the root diseases and frame canker by regular periodic application of biofetilizers and biocides at the time of pruning and incorporated.
4. Rationalize the inputs of manures and fertilizers based on the bush capacity and productivity. Integrated split doses improves the absorption and utilization efficiency.
5. Adopt ecofriendly integrated (GAP) management for plant protection, avoiding hard chemicals, liberal use of microbial products alone, and in combination with herbals extract encouraged.
6. Develop practices to manage the weeds (Integrated Management) avoiding herbicides.
7. Maintain the health of bush canopy and harvest crop periodically ensuring 32 rounds per annum.
8. Implement the cold weather operations meticulously and prepare the bushes for the next season; foliar spray of NK (urea + muriate of potash: KCl) @ 1% has been found to alleviate/mitigate the moisture stresses, maintaining the leaf water potential around 7%–8%, thereby keep the stomates open facilitating the gas exchange and photosynthesis of leaves contributing to productivity. Excess soil moisture/flooding,

inundation, severe drought/moisture stress in the soil, force the leaf/shoot water potential to 15, effecting complete closure of stomates, and shutting down gas exchange, thereby safeguarding the life of the bushes (Figure 6.2).

The Maintenance leaves play an active role on Budbreak and Flushing Behavior Rather than Roots.It has been proven in field experiments by reciprocal grafting of Assam, Cambod, china clones on a common Root stock of 80 years old seedling stock.

A Composite plant with Reciprocal grafting on a 70-80years old common Root stock Reveals that Foliage controls the flushing behavior of clones TV1,TV18,Tv20 Rather than Roots.. Thus Maintenance leaves/Canopy plays an important role on the function of the entire plant on growth, productivity and quality

Figure 6.2 Composite plants with Reciprocal grafting.

CHAPTER SEVEN

Mineral nutrition in tea

Contents

7.1 Essential nutrients

N, P, K, Ca, Mg, Zn, S,B, Mn, and Mo are considered important for the sustainable productivity of tea plantations, the quantum required being determined by many factors like cultivar, productivity, type of soil, and the management practices. In order to maintain the sustainability, rationalized, integrated nutrition management (INM) is recommended.

Being a perennial crop with intensive management practices enforced exploiting the full potential of the plantations; due care is essential in rationalizing the inputs, based on the research results, field observations, and the economics of the operations. In addition to soil application, foliar applications especially of micronutrients are done for quick results and for specific objectives. The minerals listed are categorized in to macro, secondary, and micro/trace elements depending on the quantum of requirement. Based on the field situations, cultivar, and productivity, there may be some overlaps. Role of these elements are broadly narrated in the following section.

Scientific Perspectives of Tea Plant Horticulture and Productivity
ISBN: 978-0-12-823444-0
https://doi.org/10.1016/B978-0-12-823444-0.00007-8

7.2 Macronutrients

N, P, K: macronutrients are constituents of amino acids, conservation and release of energy (ATP-ADP) through beta oxidation of fatty acids, co-factors regulating the enzymic reactions, starch metabolism, and stomatal regulation. K is not the constituent of any metabolite but freely mobile in the system. It is one of the minerals sought after by growers for sustainable productivity with quality and stress tolerance.

7.3 Secondary nutrients

Ca, S, Zn, Mg, and Mn:

1. **Ca**: Main constituent of calcium pectate contributing the building block of cell walls; besides, it is a principal soil amendment in plantations maintaining the acidic pH.
2. **S**: Constituent of amino acids like methionine, cystein, cystine, scleroids, and volatile aromatics. It serves as a quality regulator besides being an acaricide and soil amendment.
3. **Zn**: Zinc is a cofactor in cleavage of water molecule in photosynthesis, besides synthesis of auxin indoleacetic acid (IAA) through tryptophan and mediating bud dormancy in growth and development.
4. **Mg**: Magnesium is the constituent of the pigment chlorophyll, besides a cofactor of enzymes kinases on energy regulation.
5. **Mn**: Manganese is an element plays an important role in synthesis of volatiles, flavors besides substituting deficiency of magnesium in adverse soil conditions. It is suspected to play a role of enhancing the quality of tea, through synthesis of intermediates like mevalonic acid and salicylic acid.

7.4 Micronutrients

Zn, Mo, B, Si: These elements play a role in adverse soil conditions like clayey heavy soil, water logging, high productivity with excess inputs, stress situations as a moderator or ameliorant, besides maintaining quality in high productivity areas. These micro, secondary, and trace elements are required in small quantities (mg level) compared to the macro, for health and productivity of the plants. However, their deficiency or excess over the range affects productivity exhibiting characteristic symptoms on the leaves.

7.5 Some of the salient points on the nutrition of tea

NPK at 2:1:2 ratio is recommended for balanced response in health and productivity with economy, nevertheless 2:1:3 ratio is advocated, in heavy pruned (height reduction as well as rejuvenation) bushes.

Split application (3–5) is preferred for efficiency and economy.

Integrated manuring with organic bulk products is advantageous, ground application with incorporation by hoeing is recommended.

Apply manures when soil moisture is adequate, feeder roots, and canopy leaves are active and healthy for efficient absorption and assimilation.

The dose is based on bush population and productivity with a ceiling, for sustainability.

Micronutrients are need-based and foliar applied as a cocktail, in the morning or evenings; a minimum of two hours rain free period is required at least for 50% absorption.

7.6 Key symptoms of a few important micronutrients

Some of the Symptoms of common occurrence are narrated with illustrations as an aid to identify and rectify to sustain the growth and productivity of the tea bushes. Plates on def. Symptoms: N, K, Zn, Mg, B, and Mn (Figure 7.1 and Table 7.1).

Table 7.1 Nutrient deficiency symptoms of tea plants.

S. No.	Nutrients	Deficiency symptoms
1.	Nitrogen	Plants exhibit slow stunted growth, and mature foliage become pale green. Defoliation begins with lower leaves and extends to upper leaves in severe cases, the lower leaves have a "fired" appearance on the tips, turn brown, usually disintegrate, and fall off.
2.	Phosphorous	P deficiency is characterized by dull bronzing of recently matured lower leaves with or without brown spotting and loss of glossiness. The stems become shorter and finally shedding of leaves sometime leading to death of plants.
3.	Potassium	Marginal scorching or chlorosis or necrosis of matured leaves along the leaf margin. Some leaves show pronounced purple color. Bronzing over the entire leaf blade. Premature leaf fall. Development of thin and twiggy wood.

(Continued)

Table 7.1 Nutrient deficiency symptoms of tea plants.—cont'd

S. No.	Nutrients	Deficiency symptoms
4.	Magnesium	is characterized by yellowing of older leaves, and an inverted V shaped yellow tint between the veins. Inverted V shaped chlorosis. Outer half from the tip becomes pale yellow and premature leaf fall.
5.	Sulfur	General yellowing of younger leaves "net veining" in the younger leaves, leaf blade takes a striking yellow color and branching dark green.
6.	Zinc	Zinc deficiency leads to rosetting of leaves. It also leads to interveinal chlorosis of the upper (youngest) leaves. Short internodes, chlorotic, small sickle shaped leaves, wavy margin, reduction in leaf size.
7.	Boron	Corky, hooked buds, cork growth in petioles, bronze colored mature leaves.

Figure 7.1 Deficiency Symptoms.

7.7 Nutrient management

NPK are the macronutrients applied every year on the basis of expected average yield. When the production increased due to the reforms on staggered pruning, plant intensity, refinement of processing quality black teas, micronutrients like zinc, magnesium, manganese, boron, and sulfur were recommended for balanced growth and sustainable productivity, avoiding any deficiencies due to adverse interaction or toxicity, which can affect the production weakening the bushes. The NPK doses are suggested at 2:1:2 ratio in general though marginally increased potash is recommended in specific areas like red rust infestation due to water logging resulting low potash, mite-affected weak areas, or the mother bushes ear marked for taking cuttings to increase the starch reserve. The ratio of NPK and level of micronutrients are based on the analytical data of tea bushes—wild and cultivated. The levels of NPK that has been recorded in plucked shoots is 11:4:15 but varying between the cultivars and their wild parents (Chinese work published in Pakistan J of Science).

However, in the past 30 years or so, indiscriminate, excessive use of fertilizers, hard chemicals for plant protection ,and herbicides damaged irrevocably/poisoned the soil, plant, and environment to the present day situation of chronic stresses, weakened exhausted soil and bushes, have been further aggravated by the emergence of virulent strains of pests and diseases not amenable for the available routine control measures warranting repeated use of hard chemicals at higher doses impacting on the low productivity and extended intensive banjhification. The situation further eroded, resulting in poor quality, nutrient deficiencies, dieback, defoliation, vacancies of both tea and shade trees, stagnated production/yield reduction and quality deterioration, fetching the poor price realization, loss of revenue and bankrupt, of many leading plantations/corporates.

The present adverse situation and conditions are narrated below to understand the gravity and act fast to correct in an emergency basis.

1. Soil: Low pH, highly acidic, poor humus content, depleted microbial population and feeder root, starch reserve exhausted, incidence of virulent pests and diseases aggravated. The soil biosphere has been completely neglected and ignored, while the frame and canopy (plant biosphere) are overexploited exposing the plant for considerable stresses that have to be corrected, rectified, and restored for quick recovery.
2. Plant: Under hyperacidic conditions, the feeder roots get burned, and some of the divalent molecules like calcium, magnesium, zinc, B, and S become scarce and deficiency symptoms develop affecting the metabolic functions. High iron (Fe) interacts with other divalent, trivalent elements hindering the metabolism and thereby the health and vigor of the leaves. Mn builds up high in hyperacidic conditions that play a role on volatiles/

flavor quality. Any stress on leaves, biotic and abiotic, hinders the photosynthetic efficiency of leaves affecting the productivity and quality.

Under high inputs of NPK regime the soil becomes acidic fast and ends up marginal around 4.5 by the end of the first year. The plants go through the stress of suboptimal pH by second year end, the pH being around 4.1/4.2 warranting soil reclamation. Midcycle application of dolomite helps to restore the pH, boosting the performance of the canopy. Leaf addition on the plucking table in April, spring, and the fall, October, promote the quality production of second flush and the Autumn flushes giving a boost to quality and stress amelioration.

Instead of bulk quantities at a time, split applications in smaller doses improves the absorption efficiency, with less impact on soil pH. The split application is recommended besides cutting down the NPK level to 50% matching with equal quantity of organic manures like cattle manures, compost, oil cakes, and so forth to save and enrich the soil. The last split for conditioning the soil for winter can go along with soil amendments and the biofertilizers and biocides duly incorporated by cheeling/hoeing.

Thus, by judicious, rationalized, balanced split application of NPK fertilizers with organic manures at 50:50 ratio integrated with soil amendments, bioinputs, applied during winter before pruning and incorporated by cheeling/hoeing, not only helps the soil but serves as mulch, conserving soil moisture besides checking the weed growth. Being winter, the bushes get recharged with starch reserve, go through the stresses better and recover early from pruning and bud dormancy. Because of the improved health of soil and plant, the bushes go through any stress, recover fast from the bud dormancy reflecting on the early crop quantity/quality and price realization, without much problem due to red spider mite (RSM) or drought. It will be a win—win situation benefiting all stake holders of plantations, adopting the GAP early to reap the benefits.

The recommended ratio between the macronutrients NPK is 2:1:2 subject to an increase of K under special circumstances like severe incidence of mites, red rust, and so forth under chronically waterlogged situations, besides the mother bushes earmarked for taking cuttings for propagation.

As a compromise and prudence measure to revive and save the present situation, Assam, Arunachal can be declared Good Agricultural Practices (GAP)/organic, while Darjeeling and Nilgiris can maintain the Geographical Index (GI) standards restoring the productivity and quality, as well as price realizations. The small growers of Terai, Kangra, Palampur, HP, can take advantage of the soil and terrain conditions of the regions through intercrops/diversification and value addition of producing specialty teas, herbal tea, and so forth, processing and marketing the value-added products to defense forces, railways, and tourism-oriented enterprises.

As for South Indian tea, GAP should be enforced, ameliorating the soil and plant health, reducing the chemical load, fine-tuning the field package of practices and ensuring the improved production with quality through balanced, need-based rationalized inputs, mechanization, diversification with horticultural intercrops, value addition through innovative processing, blending, packaging, and online trade with contract, ensuring the cash flow and economy. Amalgamation and diversification, partnership with work force on cooperative basis, and prospective shareholding have already been demonstrated by Tata Tea Ltd. Welfare of the trained work force and dependent artisans and population is important. Alternative employment opportunities should be ensured if the plantations are closed/abandoned in the interest of the environment and wild life preservation in sanctuaries.

*The Recent tragedy of landslide in Nayamakad, Munnar, Idukki district, kerala, is an eye-opening moment showing that plantations in all hills should look toward fine-tuning their field management practices, strengthening the three-tier vegetation system of shade, intercrops, grasses, and wind barriers in the boundaries, and staggered trenches in contours against the slopes, preserving the soft herbal weeds which serve against soilwash, erosion, and landslides in future. Similarly, floods in Assam, Himach Pradesh are common factors the plantations and horticultural farms go through every year. Strengthening the vegetation in the river bed areas, stabilizing the flow of water in stable course, encouraging cleaning, and dredging to ensure surface transportation on priority will go a long way in reducing the flooding, improving the economy, growth, and employment opportunities for the upcoming younger generations. The positive impacts of these infrastructure, soil and water conservation, horticultural projects worth to be seen within two years for the welfare of people of the regions and the nation at large.

* Refers the tragedy of landslide of Idukki District Munnar Tea Plantation Kannan Devan Hill plantation, TATA Tea, Munnar "vide the Hindu report dated January 2021."

7.8 Salient points on nutrients management in tea plantations

1. NPK doses at present levels should be reduced by 50%, balancing with organic manures, composts oil cakes, Biofertilizers, soil amendments, integrated, applied in splits, and incorporated for maintaining the soil acidity, humus, microbial population, root proliferation, and maintaining the soil moisture and aeration at 60:40, adopting GAP/precision/natural farming for strengthening the soil biosphere for sustained health and productivity, free from biotic and abiotic stresses.
2. Need-based foliar application of micronized nutrients with proven organic growth boosters will take care of the deficiencies, if any, induced by interaction or stresses sustaining the growth and productivity.
3. Healthy, adequate canopy with regulated shoot density will ensure the max photosynthesis, assimilation, and metabolism coordinating the dynamic physiological parameters and reflecting on the productivity and quality, maintaining the health of the bushes.
4. Maintenance of adequate regulated shade with cover crops, herbal mulch, wind barriers, fuel baris, and so forth, sheltering the biodynamics of birds and animals will ensure no need for any hard chemicals in plantations.
5. Enforcing GAP with innovative mechanization of a few field operations such as pruning, harvesting, foliar application, brush cutter, processing, and so forth will ensure cost-effective management and productivity.

A bush sample pruned for LP/CA in November will be rested for a month and pruned in Dec.

Pruning conserved by application of BioDigester within three days of cut conserves up to 200 mt OM.

CHAPTER EIGHT

Management of tea plantations: plant protection including weed control

Contents

1. Weed control(Management)
2. Principal pests and diseases of north and South India
3. Important pests of North east and South India
4. Integrative measures suggested for the important pest and diseases

Integrated agronomical practices and the integrated pest and disease management are a two-pronged strategy for sustaining the health and productivity of plantation. Vulnerable bushes succumb to various pest and diseases and there is likely to be a crop loss of 15%—20% that is unbearable in any commercial proposition. The cultural practices and integrated pest and disease management are requires to maintain a healthy bush.

(Table 8.1: agronomical practices and impact on pest and diseases, Table 8.2: plant zones, pest and diseases and control measures).

8.1 Principal causes for pests and diseases

When the growing conditions are optimum and favorable, the pathogens and maladies are subdued. The moment stresses and unfavorable soil and climatic conditions prevail, the bushes become weaker and vulnerable,

Scientific Perspectives of Tea Plant Horticulture and Productivity
ISBN: 978-0-12-823444-0
https://doi.org/10.1016/B978-0-12-823444-0.00008-X

and then succumb to various pests and diseases. The pathogens and maladies proliferate, which make the host plant become weak and succumb to disease.

Understanding of the biology of the pests and diseases, identification of the predisposing factors, and the corrective actions with integrated pest and disease management is required for successful control of pests and diseases.

8.2 The principal pests and diseases of north and south India

Table 8.1 Important pests of North East and South India.

S.No.	Classification	Pests
1	Leaf feeding pests	Scraping and sucking pests: Thrips, red spider, purple, pink, yellow, scarlet, and pale mites **sucking pests**: tea mosquito bug, aphids, scale insects, jassids, semilooper **Biting insects or caterpillars**: Flushworm, leaf roller Nettle grubs **Chewing and biting**: Caterpillars, Tea tortrix, Caterpillar
2	Stem borers	Shot hole borer Red coffee borers
3	Soil pests	Parasitic eelworms (nematodes: Melodogynes and Pratilenchus) Root mealy bugs Cockchafer grubs Termites: live wood-eating sp. are more serious

Table 8.2 Important diseases of tea in North East and South India.

S.No.	Classification	Diseases
1	Leaf diseases	Blister blight, gray blight, black rot, brown blight, red rust, frog eye spot
2	Stem diseases	Fusarium die back, proria branch canker, thorny stem blight, Gray blight (*Pestelotia*)
3	Root diseases	Red root rot, charcoal stump rot, black root rot, brown root rot, violet root rot, diplodia disease

8.3 Integrative measures suggested for the important pest and diseases

Pest management: Integrated pest management such as cultural, chemical and biological to be adopted for management of diseases. For chemical control measures, recommendations issued by Tea Research Institutes with plant protection codes should be followed. Proven biocontrol measures for certain pests may also be followed, which are given below:

Mites: Two to three application of *Verticilium lecanii* + *Paeciliomyces fumoroseus* at fortnightly interval followed by sulfur formulations and *Azadirachtin* (neem) may be followed during this pest period.

Jassids, Thrips, Aphids: Two to three application of *Beauveria bassiana, Pseudomonas fluorescence*, fortnightly interval preferably after harvesting.

Helopeltis, Semilooper, Termites: Two to three applications of *Beauveria bassiana, Pseudomonas fluorescence*, and *Metarhizium anisopliae* at fortnightly interval. Thorough drenching of the canopy and soil is required. Hard plucking and breaking of dried stubs important in Helopeltis damaged fields.

Diseases Management: Integrated Disease management such as cultural, chemical and biological to be adopted for management of diseases. For chemical control measures, recommendations issued by Tea Research Institutes with plant protection codes should be followed. Biocontrol measures such as application of *Trichoderma viride, Trichoderma harzianum, Beauveria Bassiana, Pseudomonas fluorescence,* and *Bacillus subtiis*, by drenching spray at fortnightly intervals should be integrated.

Hail and frost damage: Foliar and frame drenching of *Trichoderma* sp, *Pseudomonas* fluorescence with micronutrients *foliar,* within three days of hail will overcome the impact. Two to three sprays at fortnightly intervals helps wound healing and hastens recovery. Addition of a maintenance leaf is suggested to compensate the defoliation/shredding of leaves if any.

In order to ease the situation and precise, biologically evolved consortiums of the straight microbial cultures have come handy to suit the natural farming, reducing the chemical loads, maintaining the health of soil and plant biospheres for sustainable productivity with quality for all maladies of soil and plants in tea plantations with advantage. Handling of the inputs also made easy by the working, force, economically viable (personal communication, Thamizh Chelvan, Vishakan, Biotech Coimbatore, 2019).

8.3.1 Integrated nutrients management (INM) and Integrated pest management (IPM) (GAP) modules developed for tea cultivation in natural farming-India

The following are four formulations of a cocktail of proven isolated strains of beneficial microbes from the soils of tea gardens of North and South India. Judicious, timely application and preparation of the spray fluid through fermenting these four bioconsortiums will take care of the entire cultivation of tea avoiding chemicals in natural farming/good agricultural practices (GAP) in an ecofriendly way, effecting sustainable productivity with quality, preserving the soil, plant biospheres, and the environment. Each of the four consortiums may require to be applied twice in a year based on the terrain climate, crop distribution, local situations, and management practices followed. Compared to the conventional practices of integrated nutrients management (INM) and integrated pest management (IPM), the ecofriendly package suggested will be economical, effective, sustainable productivity with quality, assuring the health of the soil, plant, and the working, consumer population, safeguarding the interest of the industry and the nation (Thamizhchelvan and Manivel, 2019). The bioconsortiums work and have been proven effective in other horticultural crops such as coffee, cardamom, and spices.

1. Bioconsortium soil biobooster, for quick growth and root proliferation preserving the soil biospheres
2. Bioconsortium compost, for quick digestion of prunings, dry matter, and stubbles enriching organic matter and adding humic acid for the soil, encouraging beneficial microbial build up
3. Bioconsortium rot arrest for control of all diseases of the top and roots through soil drenching as well as foliar
4. Bioconsortium biodefense for control of all the pernicious pests of tea sparing the predators

The bioconsortiums are made of the proven straight microbes with required media, nutrients, boosters, and so forth biotechnologically that are effective for the management of all pathogens and pests in tea, and an ecofriendly way for sustainable productivity with quality, preserving the health of the soil and plants without residues or side effects. As most of the microbes are stored in spore form, the consortiums can be stored and transported while keeping quality. Once they are fermented and diluted with prescribed media, nutrients, and growth boosters, the microbes multiply and release enzymes, cofactors, hormones, etc., which becomes a holistic formulation to give the expected results. (Thamizh Chelvan, 1996 PhD Thesis). Adequate care and attention taken by the management in preparation and application is ensured to get the best results in the fields.

The methodology of preparing one of the bioconsortiums is elucidated for illustration:

Soil reclamation/soil biobooster for growth and root proliferation.

8.3.1.1 Soil biobooster: plant growth promoting rhizosphere microbes (PGPR)

It is a microbial consortium of biofertilizers, biofungicides, bionematicides, bioinsecticides, and other soil mineralization and solubilization microbes. It helps soil mineralization and better absorption to the plant growth and also controls various plant diseases, pests, insects, nematodes, and so forth.

The beneficial microbes secrete phytohormones in the plant root region to facilitate quicker and vigorous growth of root and shoot.

Preparation:

Take five pkt/ha. of soil biobooster mixed 200 l water, and ferment for 10 days. After fermentation add 1000 mL of V-COMBINE and dilute to require amount of water (per hectare drench volume) and drench around root zone area of 1 hectare.

Benefits: Microbial fermented product promoting quick root growth proliferation

1. It will help to mineralize the sediment nutrients from the soil and make it available to the plants.
2. After mineralization, the soil loosens and the root proliferation improves.
3. After loosening of the soil, there is better, new root formation, and better absorption of nutrients by the plants.
4. It will improve soil structure and become healthy.
5. Because of the increase in beneficial microbial population harmful population reduced or suppressed, soil and plants become healthy. A minimum of two rounds of spray required per year. Approximate cost per Hectare is Rs.2500 II Pancha Kavia/Dasa Kavia, Panchamirtha for organic cultivation as a biogrowth promoter.

Awareness has come from the consumers and producers about the organic cultivation of the produce through integrated nutrients and plant protection measures, reducing the chemical load in horticultural crops and plantations on priority. The cow dung + urine—based Pancha Kavia formulation developed by a medical professional is being used with advantage. Being popular with growers it has been upgraded into Dhasa Kavia and further expanded to larger, corporate farms, and plantations (Natarajan, Nammalwar, Parlekar).

There are also fermented products in plantations using their own cattle population, introducing microbial strains isolated from tea gardens are multiplied and drenching the soil or foliar applied periodically with advantage for their ecofriendly cultivation of tea in Assam. In order to hasten the fermentation, additive boosters such as, jaggery, molasses, curd, ghee, and fruit like banana are used with periodical stirring for aeration with low light and for plant protection (20%) besides being the biogrowth promoter of roots and plants (80%), impacting crop increases up to 10%—15%. The concentration of spray fluids should not exceed 3%—5% to avoid scorching due to phytotoxicity.

As dairying has been proposed in plantations as diversification and value addition, Indian tea industry is poised for rapid growth and revolutionize in the near future.

8.3.2 Organic cultivation

There is awareness and demand for organic produce with least chemical residues globally. To date about 10%—17% of the food and agricultural produce are organic fetching, compensating the additional efforts and inputs required over conventional production. Thus the area and demand of organic cultivation is being encouraged, and there is enough scope for further expansion of organic cultivation of tea as value addition especially for the small tea growers, world over with advantage. As a first step the entire gardens can be declared GAP, avoiding hard chemicals, and about 10% of the area can be earmarked for organic, with adequate strategy and a plan for transition and implementation within three years. The principal package of field practices are followed avoiding the usage of hard chemicals, adopting ecofriendly integrated, indigenous herbal bioproducts while maintaining the dose/time of application for the expected crop yield under normal situations/prevailing conditions. The inputs should be produced locally available resources and self-dependence. The soil, plant, shade, status, water management, and so forth should be optimized for sustainable productivity with quality. Other relevant field operations/practices like pruning, harvesting, and foliar applications should be need-based, judiciously monitored, and implemented based on the impacts on growth and productivity response controlling the cost of production (COP). Creation of stress-free environment, soil, and plant biospheres strengthened with effective timely harvest of the crop will ensure the health and productivity, free from any maladies, pests, and diseases. Thus, a judicious, timely decision of transforming the entire garden GAP, a portion of which as organic, will pave the way for prosperity and well-being of the tea plantations especially the small growers, taking advantage of diversification and value addition.

8.3.2.1 Salient points on transformation into organic cultivation of tea with value addition are narrated for information and implementation

8.3.2.1.1 Preplanting

1. Earmark the manageable area with comfort about 50—100 ha.
2. Attend land preparation with rehabilitation, organic matter content, soil ph with lime/amendments application, aerate the soil through solarization, ensuring free from soil pathogens, termites, and nematodes.
3. Level the soil with gentle slope for draining the excess water, avoiding water stagnation/inundation, etc.
4. Mark the rows and pits for staggered double hedge leaving pathways for mechanized harvesting and foliar application, spacing for shade trees, live hedges for biomass producers, and cover crops.
5. Provide/design drains: primary, secondary, tertiary for effective drainage, ensuring optimum soil moisture ratio 60:40. Ensure staggered shallow trenches across slopes, growing vettiver/lemon grasses minimizing the soil wash, and erosion.
6. Establish tall growing, hardy fruit trees in the boundaries as wind barriers, sheltering desired flora and fauna; biodynamic environment.
7. Plan intercrop with locally popular established ecofriendly fruits, spices, and condiments, herbs.
8 Border crops like French Marygold, Pyrithrums, shrubs such as Lantana, Adathoda Vitex negundo, Seehapal (*Annona squamosa*) and similar deterrents can be raised as trap crop.
9. Principal recommended horticultural crops regions:
    1. Assam: Mandarins, pineapple, ginger, turmeric, graviola, capsicum, cinnamon, pepper, nutmeg, allspice anato dye, citronella, and vettiver grasses.
    2. Darjeeling: Mandarins, pineapple, peaches, olive, plum, apple, apricot, avocado, saffron, and stevia.
    3. Terai, Himachal: Apple, pear, plum, peaches, litchis, persimmon, olive, jamoon, saffron, and anonas.
10. Form cooperative production companies under precision farming, pool, organization, warehouses, processing, packaging, and export.

8.3.2.1.2 Planting and postplanting

The planting pits are filled with recommended ingredients like Trichoderma, Pseudomonas, Metarrhizium, Phosphobacter, vermicompost, neem cake, rock phosphate, and so forth well mixed and filled up to the brim, ramped, and staked against wind. Irrigated to encourage the setting, removal of air pockets, if any, and ensuring root contact with soil colloids. The space between interrows are mulched, sparing the collar zone to conserve the soil moisture with aeration. Simultaneously leguminous cover crops, planting of temporary shade trees, Tephrosia, Crotolaria, Sesbanias, Glyricidia, and Indigofera can be raised for shelter of young plants with humidification and nitrogen fixation, and promoting the growth of young tea plants. Adding of any chemical pesticides in the pit or usage of herbicides, systemic chemicals spraying are prohibited in young tea for organic cultivation or GAP. Thus, the young tea of some good quality, productive cultivars like P126, Av 2, S3A3 TRI 2045, TRF4, DP1, and TV 22/23, can be utilized both for black and green tea with advantage depending on the terrain, climate, and rainfall distribution. Strengthening and standardizing the soil, plant parameters/biospheres balancing the nutrient inputs, regulated shade thereby the microclimate, free from pernicious pests and diseases are bound to reflect on the clean crop with quality will improve the economy and prosperity of small growers, corporate, adopting organic, precision, natural farming, and GAP.

With the present-day technology, available inputs, transformation, and adoption of organic cultivation, coupled with diversification and value addition are expected to be abundantly successful.

8.3.2.1.3 Mature tea-new/old plantations: organic tea cultivation/package of practices, salient points

1. Optimize the soil, plant, shade, and environment parameters for sustained productivity with quality.
2. Estimate the expected yield with zero budget rationalizing the need-based inputs like NPK biobased micronutrients, biostimulants, 100 kg N at 50:50, organic/inorganic mobilizing all possible resources of the garden, split doses, and foliar feeding, hoed, and incorporated removing the weed.
3. Plan the crop distribution and organize the work force with appliances,harvest the crop effectively with 65% fine. Shear harvest preferred by trained workers.

4. Foliar feed the bushes with nanomicronutrients with sea weed extract (amino acids, etc.) biostimulants, PGPR/herbal extracts after every second/third round of harvesting for sustaining the health of bushes.
5. These foliar sprays can be combined with appropriate biocides/bioconsortiums for taking care of the maladies, if any (INM, GAP recommended packages), based on need.
6. Field proven and recommended/suggested biocides given below: the dose will vary depending on the local fields situations, used alone or in combination; bioconsortiums are broad spectrum, cost-effective during peak crop, timely, and labor saving.

   6.I Bioconsortium:

   Maladies: Principally pests, diseases, physiological disorders.

   Biocides: Alone and combination "Bioconsortiums" for effectiveness.

   Soil Reclamation Biocides: Trichodermas, Pseudomonas, Metarrizium

   Biofertilization: Azspirillum, Phosphobacter

   K+ solubilizer/mobilizer and vesicular arbusicular mycorrhiza (VAM) Bio-Booster fermented with starter for three days, diluted and applied with drenching soil and foliage with jiggery/vermin concentrate February/March—early crop applied fortnightly with a minimum two rounds per year.

   Root foliage diseases BV, *B. subtilis*, Pseudomonas, *Trichoderma viride*.

   Panchakavia, DasaKavia, and herbal extract alone, and combinations of rot arrest during April/May monsoon peak. Applied fortnightly twice a year.

   Foliar frame pests/diseases Trichoderma viride/bassiana, Verticilium, Paecilomyces, cow dung–urine herbal extract, and biodefense June–Sept peak crop, applied fortnightly two rounds per year.

   Biomass conservation mushroom spawn, Aspergillus niger Biodigester Pruned bushes, prunings

   6.II Protocol for tea, horticultural crops for Assam/Bengal, North East, and organic/GAP cultivation—bioinputs

   No bioproducts and usage method of application time of application in plantation dosage.

No.	Bio-Product	Usage	Time of application	Dosage
I	Trichoderma	Bio fungicides	April/October	2-5 Kg Per Ha.
II	Azospirillium	Bio fertilizers	Planting time	25 kg Per Ha.
III	Phosphobacteria	Bio fertilizers	Planting time	25 kg Per Ha.
IV	Metarrhizium	Anti termites	Against Termites	5 kg Per Ha.
V	Beauveria	Bio-fungicide	Antilooper, Fly Maggots, Borer	5 kg Per Ha.
VI	Bio Digester	spray on fresh pruning's	December/January	5 kg Per Ha.

7. Maintain a separate register for any inputs, source, and dose applied, facilitating traceability.
8. Maintain the stock of inputs properly organized for inspection by certifying agencies, assuring the availability and timely application.
9. Suggested to use conserved rainwater or clean, running stream water with need-based use of nonionic sticker/spreader during inclement monsoon weather.
10. The spraying crew should be provided/protected with masks, sunglasses, aprons, and gloves. All appliances, including, harvesters, and spray appliances should be clean, checked periodically maintained without down time. First aid kits should be provided with the supervisors while the field operations.

TRF4 an elite china Assam Hybrid clone suitable for High Ranges with quality orthodox black tea.

6.2 oilcakes, poultry manure for P, wood ash K solubilizers, sea weed extracts, mineral extract of potassium ores for K. Thus the level of NPK could be reduced with an integrated approach, sustaining the productivity with quality for economy. Intercropping with crops like mandarins, avocado, durian, mango, cinnamon, cloves, nutmeg, allspice, and cardamom is encouraged for value addition, environment protection, and economy.

Plant Protection:

The occurrences of pests and diseases are mainly governed by environmental and other soil factors including bush health and field sanitation. A healthy bush is not affected by either pest or diseases. Shade regulation and optimization of soil acidity, are important. Bush health and productivity depend on the health and volume of roots and canopy.

Helopeltis and red spiders are the main pests while the blister blight and gray blight are the common diseases. Suggested management measures:

Mites: Verticillium + Paeciomyces combined spray, two rounds fortnightly. Micronized sulfur, neem kernel extract, cow dung urine extract etc can also be tried intermittently.

Helopeltis (Tea Mosquito): Beauveria + Pseudomonas cocktail fortnightly. Aphids, Thrips and Hoppers (sucking insects): Integrated measures employing bio (BV, Pseudomonas), herbal (neem, pungam, Artemisia, Calamus), and cow dung—urine extract at low doses. Once the predators are encouraged by avoiding hard chemicals and stresses are mitigated, these sucking insects will not be a problem.

Termites: Metarrhizium with jaggery at the time of pruning.

The prunings should be saved either by burying or treatment with biodigester.

Gray Blight (Pestalotia): Trichoderma + Pseudomonas. Foliar application of nano bioproducts with amino acids and micronutrients, after every third round of shear/mechanical harvesting. It saves the foliar of micronutrients also.

Blister Blight: Bordeaux mix/Copper Oxy Chlorides, restricted to 15 rounds per year.

Frame Canker: (Pomopsis/Poria) Trichoderma + Pseudomonas + Bacillus subtilis with molasses at the time of pruning.

Besides these microbial products, herbal extracts, cocktails, neem and Pungam-based products, and field management practices can be used with advantage.

Points to remember for ecofriendly cultivation:

1 .Nutrients are one of the factors for crop yield but not the only factor to be rationalized.
2. Once the soil pH, OM level, shade status, bush health, and sanitation are ensured, plant protection is taken care of, productivity with quality ensured.
3. At least integrated approach for GAP is ensured in the beginning, so that organic cultivation/diversification with value addition are followed within a set time frame.
4. Discretionary powers for use of chemicals, if any, the manager on the spot can use his discretion to face any emergencies, provided they are traceable and notified, in records, avoiding crop loss.
5. Thus the south Indian tea plantations have to take up many field management and manufacturing processes on priority, Restoring the soil parameters, bush health, and manufacturing procedures, aggressive

forward contract marketing with assured export of quality tea, ensuring cash flow. Illegal distribution of substandard, adulterated tea in retail outlets should be controlled by the Food Safety and Security Act (FSSAI) and enforced strictly.

Rationalization of inputs with integration with OM, vermin compost, oil cakes, consolidation of shade and tea, liberal use of biofertilizers, biocides in management of pests and diseases, stress management, fine-tuning pruning, harvesting of crops, manufacture of black and green tea with value addition, reducing the chemical load, preserving the water resources, soil, and bush health building up the environment, ensuring quality control of market tea are the priority areas for the UPASI to act, take control of need-based research projects with dedicated accountable team with responsibility. The future of the tea industry is bright.

Need-based use of proven, ecofriendly growth promoters for young tea management, bush architecture, pruning, harvesting, micronized micronutrients with seaweed extract, amino acids for crop distribution, stress amelioration, NK foliar, biocides both bioconsortium and straight biocides in combination with herbal fermented products like Pancha, Dhasa Kavias may very well be used with advantage. The required package of practices should be developed and recommended by the R and D UPASI for adoption with confidence.

The dynamic eco balance between flora and fauna both ambient and soil biospheres are maintained by nature for the benefits of all living creatures on earth. Any disturbance/interference on this delicate balance impacts on creation of many adversaries, particularly on agriculture, horticulture, plantation and spice crops irrevocable maladies, pests and diseases, natural calamities. Our intensive cultural practices aiming food security and safety in the past 70 years overexploited the resources and damaged the ecobalance adversely. Most of the beneficial predators and parasites disappear or the population is reduced, giving way to the emergence of virulent pests and diseases not amenable to the inputs or packages available.

Time has come to restore and reverse the situation conducive for growth and development, buildup of ecofriendly beneficial predators, parasites, bioorganisms—microbes, fungal, and bacterial, so that a balanced soil and plant biospheres are established through ecofriendly GAP, precision farming, integrated nutrients, field management practices for sustainable productivity with quality ensuring the longevity/life span of our tea and horticultural crops with advantage. Afforestation, environmental upgradation, dynamic biosystems, Reduction of Chemical loads are the priority areas for all the stake holders involved in the interest of the society and nation.

Increased, deforestation, degradation of soil and water resources, consequential land slides and flooding, migration of trained workforce to plains deserting plantations are envisaged as reality in near future. Welfare measures and resettlement of the workforce, their housing, education of children, skill development centers, as shareholders in plantations ensuring their future will pave the way with harmony, egoism, and education. Skilled and emerging society of owners, management, workers, and consumers are involved in the well-being of all dependent population for social congenial environment for growth and prosperity people and the nation.

The future of the south Indian tea plantations established by the pioneers are in good hands for continued prosperity. The technology is advanced. The new generation should carry forward the industry with innovations, soil reclamation, refinement of field management and processing practices in collaboration with the management and working force, practicing GAP, diversification and value addition, aggressive online marketing and so forth.

A Predator spider in action against Helopeltis, red spider mite, in Assam Organic garden (Boisahabi)

Biocontrol of Helopeltis and Hyposidra, Thrips, Aphids, Jassids Resurgence in tea plantations of Doars, Darjeeling West Bengal, in progress.

Calendar operation from January to December may be maintain.

General field practices to be followed during the year.

January

LAND PREPARATION: Fill up the pits and old subsidiary drains. Complete land profile survey. Avoid plowing in locations marked for main and submain drain. Ameliorate the soil based on soil test. Follow a sequence of cross plowing—cross harrowing—subsoiling—cross harrowing—leveling.

SOIL REHABILITATION: *Mimosa invisa* seeds are to be drilled in alternate rows of Guatemala to be planted.

TEA NURSERY: Prepare the beds. Water the Autumn-propagated cuttings judiciously. Complete sowing of seeds and cover with thin mulch.

SHADE NURSERY: Apply organic manure @2.5 kg, 200g Dolomite and 125g SSP per m^2 during bed preparation. Cover the beds with fast degradable mulching material. Make the drains 60 cm deep in between the beds and 90 cm in the periphery.

YOUNG TEA PRUNING: Do the operation subject to satisfactory level of root starch and soil moisture.

MATURE TEA PRUNING/SKIFFING: Medium prune: Complete by middle of the month. Light prune and deep skiff. Complete by end of the month. Medium skiff: Complete by end of the month. Light skiff and level of skiff: Complete by middle of the month and avoid in droughty areas. Cleaning: Complete hand debanjhi and KCO.

MANURING: Continue foliar application of MOP and magnesium.

PLUCKING: Level the plucking surface and pluck off banjhis.

MULCHING: Mulch all newly planted areas.

WEED CONTROL and CULTIVATION: Uproot the ferns, creepers, and hardy woody perennials. Fork/light hoe after LP/MP in heavy textured soil. Complete ground leveling where necessary.

PEST and DISEASE CONTROL: Control on spot the initial infestation of pests. Fork the soil to collect chrysalids of looper and bunch caterpillar. Give alkaline wash on bush frame and control termite.

Bush sanitation and cold weather operations: break back the dead twigs and level up. Foliar spray of MICROSUL + NEEMFOL against cold weather pests; irrigate @ 15days interval YT, UP fields NKMg (green label magnesium) foliar against drought, spray Trichoderma + Pseudomonas against Pestalotia. Complete pruning, lung pruning preferred. Cold weather Operations: YT-deep hoe and incorporate cattle manure, hand weed and mulch, raise cover crops. Clean the drains, restore outfall, attend seed baries. Maintain nurseries. Apply MOP + GHOM 300 mL/ha for drought resistance. GHOM chelates MOP.

February.

LAND PREPARATION: Lay the subsidiary, submain, and main drains. Complete final ground leveling.

REHABILITATION: Plant Guatemala cuttings in nondrought areas.

INTERCROPPING: Sow crotalaria seeds in between alternate tea rows in proposed as well as newly planted areas.

NURSERY: Complete bed preparation, sleeve filling, and overhead shade. Regularly water the sleeves allowing the soil to settle. The depression in the sleeves to be filled up.

TEA PLANTING: Start planting toward the later part of the month subject to availability of rain/irrigation.

SHADE PLANTING: Plant shade saplings following rains. Ensure proper ramming during planting.

PRUNING/SKIFFING: Complete frame forming prune by middle of the month. Thumb prune/de-centre the established tea plant and infills. De-bud the weaker (single stemmer) plants. Complete re-centering of strong branches.

MANURING: Spray MOP and magnesium sulfate if drought prevails. Can be cocktailed with acaricides.

PLUCKING: Regularly remove banjhi from the table. Maintain a uniform plucking surface. Allow the pockets in the plucking table to fill up.

MULCHING: Mulch the newly planted areas.

DRAINAGE: Complete cleaning, deepening, regrading of the drains.

PEST and DISEASE CONTROL: Monitor incidence of pests and spray on spot. Helopeltis infested patches should be plucked black ensuring thorough removal of infested shoots. Hand collect/light trap the moths of looper and red slug. Undertake insecticidal spraying in seed bari. Complete control measures against termites.

Acaricide spray, organize and produce inputs, cheel and weed the UP and DS fields. Repeat NKMg (green liquid magnesium) + Microsul foliar, spray Karanjifol/Neemfol against sucking insects and cold weather pests. Continue irrigation to promote bud break and spray Microsul + Neemfol 5% combination for multiple pest. Spray Naturalis for looper. Spray Sucker Stopper for green flies, mites, and so forth. Or give a cocktail spray of Beauveria + Verticillium along with jaggery or molasses against sucking insects.

March. Tea plants are propagated through seeds as well as vegetation. Of late micropropagation is also practiced. TS462 and 520 are popular stocks exploiting the hybrids vigor. Seeds are stored in charcoal/sand media before raising in nursery beds.

Biclonal seed stocks TS462 and TS520 seeds are popular in Assam plantations. The trees, capsules, and seeds are shown in the three pictures.

Seeds of TS462 (S3A1xTV1): A quality, high yielding stock for black tea/orthodox ad curl twist crush (CTC).

Single leaf cuttings are taken from the mother bushes of elite clones, nurtured with special care on nutrients and stress management without tipping. The shoots are cut/harvested twice in a year. The top stem and bottom of the shoot with three leaves each are rejected. The brownish stem with mature leaves with adequate starch reserve, the buds being dormant are selected, slanting cuts are given exposing the vascular system, dipped in IBA rooting media, and planted in nursery, maintaining the natural orientation.

Propagate cuttings from ready primaries. Transfer the rooted cuttings from bed to sleeve. Water the sleeves judiciously.

Sow the seeds and cover with a thin layer of mulch. Keep the sleeves/beds moist. Complete infilling of shade in mature tea.

PLANTING: Do planting/infilling subject to satisfactory level of soil moisture. Mark out the space for planting permanent shade.

MULCHING: Mulch newly planted areas.

MANURING: Apply first split of manure in UP teas under adequate soil moisture. Apply YTD mixture in young tea. Apply YTD and sand mixture (1:9) in sleeves with cuttings of four to five leaves.

PLUCKING: Continue janam plucking of UP teas at shorter intervals. Adopt liberal plucking in teas defoliated/died back.

REHABILITATION and GREEN CROP: Complete planting of rehabilitation crop. Sow the seeds of crotalaria in newly planted tea.

WEED CONTROL: Apply preemergent herbicides if programmed? Apply postemergent weedicides like Oxyfluorfen, depending on weed growth.

CULTIVATION: Fork/light hoe LP/MP teas in heavy textured soil of droughty areas.

PEST and DISEASE CONTROL: Monitor pest incidence and control on spot. Neem or other botanical pesticides and biocontrol agents should be applied. Spray insecticides on shade up to manageable height. Remove second year green crop. Hand collect caterpillar pest and light trap the moths. Spray COC in red rust/black rot infected areas. Spray pesticide in seed bari.

Pluck liberally to level, and give a spray of GHOM + phenolic zinc (green label zinc) to promote growth. First split of NPK ground application, repeat NKMg (green label magnesium) + Microsul (fertigate if possible), Tip the LP/DS fields, Give a spray of SUPER-TE for waterlogged areas, Break the dried twigs and remove ectoparasites followed by a SUCO or COC cocktails with Microsul foliar for seed baries. Spray Naturalis in multiple pest areas. Red Spider Mite prone area use Bio-Vector, like Vertcillium + Paecilomyces with jiggery/molasses.

April

NURSERY: Continue propagation of cuttings. Complete preparation of beds for Autumn and keep under mulch cover. Remove the weeds and fill up if depressions exists with sandy soil complete sowing of shade tree seeds. Ensure that the sleeves/beds where the Tea seeds were sown do not go dry.

Care and upkeep of mother bushes, preparing cuttings for vegetative propagation.

PLANTING: Continue planting and infilling. Complete shade planting. Ensure proper ramming while planting.

MANURING: Apply YTD and sand mixture to the sleeves with plants of four to five leaves. Apply first dose of NPK manuring in young and mature teas. Apply urea and zinc sulfate as foliar on weaker teas.

PLUCKING: Pluck close to janam and remove banjhis from UP teas. Tip the DS and LP teas at appropriate measure at shorter intervals. Raise the leaf in debilitated unpruned teas toward the end of the first flush. Unpruned teas suffered from drought should be plucked liberally.

REHABILITATION and GREEN CROP: Complete planting of rehabilitation and green crop.

DRAINAGE: Clean the drains by removing the silt.

WEED CONTROL: Apply weedicides on active weed growth depending on type of weed. Avoid application of weedicides in 0 and + 1 year tea. If unavoidable spray using protective shield. Sickle between the rows and hand weed the collar in young tea. The regrowth of woody perennial and creepers should be uprooted manually.

PEST and DISEASE CONTROL: As far as possible restrict use of pesticides to spot. Use only TRA recommended pesticides at appropriate dilution on target. Restrict use of hard pesticide the safeguard the predators. Apply pesticides on shade tree trunk against Borer and Looper. Spray copper fungicides against red rust and black rot.

Second split of NPK, first round of herbicides Ghom + phenolic chelated zinc (green label zinc) for early crop, Neemfol + Microsul against sucking insects, tipping pruned fields, center the YT, give a spray of Ghom to promote lateral growth, Apply humic acid granules (Humisoil) in furrows, mulch YT, manure seed baries, SUCO spray, Formative prune YT. Spray CUO + NAA for wound healing for hail damage. Spray Bio-Vector in mite-prone areas and Rekover for early hail recovery.

May

NURSERY: Continue planting of cuttings. Repair the drip damage. Thin out the green crop from seed nursery. Spray pesticides against pests.

Suggested Planting materials for New Plantings, Crop Improvement, Diversification, Value Addition employing latest technologies, Genetcs Molecular Biology, Genetic Engineering, Biotechnology for the bright Future: Breeding, Diversification, Value addition, Mechanization/Automation/

Innovations/space and Manpower Saving, Improving the Economy of Field Management, Processing operations.

Cultivars/clones:TV1,TV 9,TV23,TeenAli17,P126,S3E3,AV2,-TRI2024,TR2025,2045, Kenya Chinary clone1.

TS 462,TS 520,Upasi BSS1, TRF4.

These have proven potential for vigorous, growth, Productivity, Stress Tolerance etc worthy of exploitation. The labs with Infrastructures and Microbiology, Biotechnology expertise, Manpower, Deserve to be built up Revving and upgrading the Tocklai, TRI with international collaboration tying up with IIT Guwahati and IIT Khragpur for the envisaged bright future of the tea industry.

Nursery plants in sleeves become ready for planting in 10—12 months.
PLANTING: Continue planting.

MANURING: Apply Urea (1%) and Zinc sulfate (1%) in plantation having banjhi problem. Complete application of first dose of manuring in young and mature tea.

Y.T. MANAGEMENT: Decenter/thumb prune the established plants. Recenter the strong central branches with 30 cm.

PLUCKING: Remove banjhi and maintain a flat plucking surface. Do not allow creep in LP/DS Teas. Tip the MP/RP teas at 30 cm above the pruning height.

REHABILITATION and GREEN CROP: Do weeding if necessary. Green crop should be thinned out. Apply NPK mixture (2:1:2) containing 60 kg N, 30 kg P205 and 60 kg K20 per hectare in Guatemala.

WEED CONTROL: Complete first round of weedicide application. Sickle the weeds in between the rows and hand clean around collar in young teas.

PESTS and DISEASES CONTROL: Restrict use of chemical pesticides to spot. For blanket application use Botanicals including neem, sulfur and biocontrol agents. Spray copper fungicides against red rust, black rot and nursery diseases. Control Fusarium dieback with two rounds of Hexaconazole (1:1000HV) at fortnightly interval.

Complete second split of NPK ground application. Add a leaf wherever the canopy is weak, complete tipping and form table. Give a foliar of GHOM + GL ZINC on tipped fields breaking interflush dormancy. Rainwater harvest and conservation, and check dams and ponds. Spray Green Cal_Bo in hail affected fields. Spray Naturalis in Helopeltis, Mites prone areas. Apply Humicel in YT/Nursery for root promotion. Spray SUCCER

STOPPER for Mite and Helopeltis prone areas. Mix 2 gm. HUMSOIL in each Nursery Sleave for better root system.

June.

NURSERY: Continue planting of cuttings. Repair drip damage. Thin out the shade seedlings and fill up the vacancies. Keep the seeded beds weed free. Adjust overhead shade to maintain adequate slope.

PLANTING: Complete planting of tea and infills.

PRUNING and SKIFFING: Thumb prune/decenter the established tea plant and infills. Debud the weaker (single stemmer) plants. Complete recentering of strong branches.

PLUCKING: Regularly remove banjhi from the table. Maintain a uniform plucking surface. Allow the pockets in the plucking table to fill up.

MANURING: Apply second dose of YTD mixture in young tea and nursery. In mature unpruned teas apply second dose of NPK fertilizer. In weak and waterlogged teas, apply urea, DAP and MOP as foliar.

WEED CONTROL: Apply second round of weedicide as per need. Resort to sickling and hand-weeding of collar region in very young tea. Avoid damage to lower branches and collar depression from manual weeding. Suggested integrated measures for weed management, sparing the soft herbal weeds and small grasses under the canopy, as mulch conserving soil moisture, encouraging buildup of microbial population.

PESTS and DISEASES CONTROL: Monitor pest incidence and adopt control measures on spot. Ensure proper supervision while spraying. Hand collect the late instar caterpillars before spraying pesticide. Avoid repetition of same pesticide. Use fungicides against red rust and black rot. Improve ventilation in Helopeltis infested areas by lopping lower branches of shade and side branches of tea bush.

Third split of NPK ground application, second round of herbicides for weed control, Neemfol or Sucker Stopper + Nutrastick. Foliar against Helopeltis, ground application of NPK for seed baries and give a foliar of GHOM + GL ZINC to promote flowering. Shade correction-lop the lower branches and regulate shade. Give a round of SUCO/BM against black rot under shade trees. Spray of Trichoderma + Pseudomonas can also be tried with advantage. Break the dried stubs and Spray Green Cal_Bo for recovery from Helopeltis damaged bushes.

July

NURSERY: Repair drip damaged sleeves. Thin out the green crop in between the beds from seed nursery. Give a round of weeding in shade nursery.

PRUNING and SKIFFING: Deep skiff the mother bushes. Thumb prune/decenter if deferred. Go for an airy skiff in UP teas with uneven plucking surface. Introduce MS in some weaker UP teas.

MANURING: Apply second dose of fertilizer in LP/DS teas. In waterlogged areas defer application to September. Apply foliar NPK mixture in waterlogged and weaker teas.

PLUCKING: Pluck at regular shorter intervals. Maintain uniformity of plucking surface. Pluck black in the event of excessive banjhi formation.

REHABILITATION and GREEN CROP: Lop the well grown Guatemala grass at 30 cm. Lop periodically Mimosa in a mixed stand. Lop the lower branches of crotalaria. Plow in and incorporate *Tephrosia purpurea* annual shrubs with advantage in plateau/River valleys.

DRAINAGE: Ensure free flow of drains. Record water table build up in waterlogged areas. Record the average flood level in outlets.

WEED CONTROL: Spray weedicide on spot depending on weed flora. Sickle the weeds in very young tea avoiding injury to the plants.

PEST and DISEASE CONTROL: Monitor and apply pesticides on spot at the initial infestation. Ineffective pesticide should not be repeated. Lop the lower branches of tea and shade in Helopeltis infested areas. Proper supervision from mixing of pesticides through its application is required. Fungicides should be applied in red rust and black rot infected teas.

Effective harvest of peak crop, Augment the capacity of factory, Measures against Helopeltis. Clear the blockages and waterways for effective outflow. A round of SUCO/BM against blister and black rot in prone areas (spot application). Rainwater harvest and conservation ponds, check dams. Add Nutrastick (Nonionic Organic Sticker) with Pesticides for better rain fastness and spread.

August

NURSERY: Repair drip damage. Hand weeding in tea and shade nursery. Thin out the overhead shade if necessary. Scrap the mossy growth and fork lightly the sleeves. Remove green crop from seed nursery. Spray appropriate insecticides as and when required.

YOUNG TEA: Check and fill up the collar depression.

MANURING: Apply a round of YTD and sand mixture (1:9) in the sleeve. Apply YTD mixture at recommended dose and method in young tea. Apply NPK manure to deep skiffed mother bushes. Apply foliar NPK mixture to waterlogged teas.

PLUCKING: Avoid creep and maintain a flat plucking surface. Pluck at shorter interval leaving smaller shoots.

REHABILITATION and GREEN CROP: Lop the well grown Guatemala grass at 30 cm. Lop Mimosa in a mixture stand if situation arise. Lop the lower branches of crotalaria. Incorporate *Tephrosia pupurea.*

DRAINAGE: Ensure free flow of drains.

WEED CONTROL: Repeat application of weedicide where necessary. Avoid spraying weedicides on drain edge. Repeat sickling and hand weeding of collar in very young tea.

PEST and DISEASE CONTROL: Use acaricides and insecticides only on infested patches. Adopt cultural control measures side by side with the chemical control. Hand collect or light trap moths of looper. Spray two rounds of COC/Hexaconazole at 15 days interval against black rot. Skiff the severely Helopeltis infested patches before pesticide application. Improve ventilation by lopping lower branches of shade and tea bush.

DAP, KNO3 foliar, GHOM + TRIA foliar to compensate light limited conditions, Sucker Stopper spray against Helopeltis, SUCO/BM against black rot, clear the blockages of drains. Effective harvest and augmentation of factory capacity. Add GHOM with fungicide spray like COC/BM. Cocktail COC with MICROSUL

September

NURSERY: To harden the plants for Autumn and spring planting thin out the overhead shade. Keep the shade and seed nursery free from weeds.

PLUCKING: Fill up the pockets by plucking up to the average level of creep. Introduce black plucking in the event of excessive banjhiness. Contain creep in LP/DS teas with the permissible limit. Allow the peripheral shoots to come up to the level of the table.

PRUNING: Lop the side branches of MP/RP bushes if vacancy infilling was done.

MANURING: Apply second split of NPK fertilizer in waterlogged mature tea. Where third split is practiced should be applied. Apply a round of NPK (2:1:2) mixture @ 30 kg N/ha in young tea. Apply additional dose of 60 kg P205/ha in tea proposed for MP/RP.

REHABILITATION and GREEN CROP: Lop Guatemala at 30 cm above the last cut mark. Cut the rehabilitation and cover crop from the areas to be planted.

MULCHING: Collect and apply mulch material in young tea.

WEED CONTROL: Apply weedicide as per need. Hand weed the young tea areas. Sickle the weeds at ground level to restrict flowering.

PEST and DISEASE CONTROL: Apply acaricides thoroughly as chances of resurgence of mite is more. Assess the level of Helopeltis infestation to adopt proper control strategies.

Apply Biocides- Trichoderma + Pseudomonas, COC/BM or Hexaconazole depending on the persistence of black rot. Spray appropriate termiticide, Metarerizium, in young tea. Spray insecticides against caterpillar and borer in young shade trees. Undertake insecticidal spray in seed baries. Hand collect or light trap the moths of looper and Red slug.

Add a leaf during Puja rush in UP, DS fields, Apply GHOM + GL ZINC foliar for backend crop. Complete fourth split of ground application. A round of Trichoderma + Pseudomonas, SUCO or MICROSUL with COC/BM against black rot and blister. A round of LIQUIBOR + MICROSUL in water logged areas.

October.

LAND PREPARATION: Use Jungle Jim for pulverizing Guatemala. Minor depression should be filled up.

NURSERY: Start Autumn propagation of cuttings. Allow the ready plants to be hardened after sorting out. Start filling up the sleeves for seed nursery. Hand weed the shade nursery.

PLANTING: Start Autumn planting of tea.

MANURING: Apply last dose of YTD mixture in sleeve plants. Apply foliar urea and zinc sulfate 1% solution at fortnightly interval. Complete application of third split if programmed.

PLUCKING: Maintain plucking round particularly in proposed UP teas. Step up by a leaf in the proposed UP teas if status of maintenance foliage is inadequate. Rest the bushes severely damaged by black rot as well as the teas to be medium pruned.

MULCHING: Mulch the newly planted tea.

WEED CONTROL: Hand weed the creepers and woody perennials. Only spot application is required. Sickle the weeds where manual control is practiced.

PESTS and DISEASES CONTROL: Spray pesticides on spot. Apply biopesticides like *Beauveria bassiana* against Helopeltis and Bacillus subtilis against black rot at fortnightly interval. Apply Metarrhizium/termiticide in the plantation having terrible infestation. Continue application of acaricide being vulnerable to mite attack.

Complete leaf addition forming a new table. A round of Green Cal_Bo in Helopeltis affected areas for recovery. A round of Beauveria + Paecilomyces + jaggery or Neemfol 5% against Helopeltis. A round of MICROSUL to wipe residual pest before Puja Flush.

November.

LAND PREPARATION: Uproot tea and shade. Fill up the pits formed uprooted tea bush and shade tree. Fill up the subsidiary and sub main drains.

NURSERY: Complete Autumn planting of cuttings. Water the sleeves/ rooting bed with hand sprayer. Sow the crack seeds in the sleeves/beds and cover with a fast degradable mulch.

PLANTING: Complete planting of tea.

PRUNING and SKIFFING: Start pruning of mature teas in nondrought areas from mid of the month retaining a healthy breather and knife clean the bush frame. Prune clonal mother bushes and knife clean thoroughly. Apply alkaline wash or Trichoderma 5% suspension on the bush frame.

MANURING: Apply foliar MOP + urea at fortnightly interval in all proposed UP teas. Spray MOP + Urea (2%) two rounds at fortnightly interval in proposed MP areas.

PLUCKING: Complete stepping up of the table wherever necessary to keep unpruned. Pluck off the banjhis from the table regularly. Keep under rest the proposed medium prune teas and black rot infected bushes.

MULCHING: Mulch the newly planted teas.

PESTS and DISEASES CONTROL: Restrict use of acaricides and insecticides only to the infested patches. For blanket application use only neem or other botanical products and sulfur formulation. Apply biopesticide like *Beauveria bassiana* against Helopeltis and Bacillus subtilis against black rot if not applied in the previous month. Treat the black rot infected areas with two fortnightly rounds of Carboxin in absence of biocides. All Helopeltis punched shoots should be plucked off.

Fifth split of ground application with oil cakes/cattle manure, with dolomite as soil amendment and incorporate by hoeing, and burying the weeds if any. A round of GHOM + GREEN LABEL MAGNESIUM for blackened crop restoring health of bushes. A round of KARANJFOL against cold weather pests. Rest the bushes due for pruning. Ensure starch reserve buildup. A round of MICROSUL in fields going UP next year to wipe residual pest.

December

LAND PREPARATION: Mark the location of submain and main drains to avoid plowing. Soil should be tested for necessary amelioration. Follow a sequence of cross ploughing—cross harrowing—sub soiling—cross harrowing—leveling.

NURSERY: Do watering as and when necessary in both tea and shade nursery. Continue sowing of tea seeds and cover the sleeve or bed with a thin layer of dry mulch.

LIGHT PRUNE: Continue pruning and knife cleaning operation.

MEDIUM PRUNE: Start the operation from mid of the month and complete preferably by the end of the month. Apply Indopaste or Trichoderma 20% paste on cut surfaces quickly after prune. Leave a healthy breather in each bush.

DEEP SKIFF: Start the operation from middle of the month.

PRUNING (MOTHER BUSH): Prune mother bushes for spring propagation. Thoroughly knife cleaning is essential.

BUSH SANITATION: Complete hand debanjhi and knife cleaning out operations in already pruned and skiffed areas. Apply alkaline wash or Trichoderma biocide (5%) suspension on the bush frame.

MANURING: Continue foliar application of MOP and magnesium sulfate. Collect soil samples for analysis to prepare manuring program.

PLUCKING: Maintain a flat plucking surface. Bushes suffering from moisture stress should not be plucked and rested.

DRAINAGE: Deepen and regrade the drains. Complete contour survey of areas to be planted in next spring.

PEST and DISEASE CONTROL: Only spot control if noticed. Take control measure for termite removing the earth runs and dead branches. Collect Chrysalides of bunch and red slug caterpillars from dried leaves on the ground as well as from fork of the bush. Hand collect or light trap the moths of Red slug.

A round of Tricho + Pseudomonas against Pestalotia. A round of GHOM + GREEN LABEL MAGNESIUM for fields entering UP. A round of MICROSUL + NEEMFOL against mites. Rest the bushes at least for a month to enrich reserve. Apply dolomite one ton per ha to correct pH. Apply Bio-Vector in mite infested fields.

CHAPTER NINE

Tea processing and quality improvement

Contents

9.1 Introduction

Tea leaves are unique, containing plenty of polyphenols, flavonoids, secondary metabolites, qualifying as a rich source of antioxidants, and ending up with tea as a health beverage. The polyphenol content is maximum in the growing tip, declining from the top downwards. The harvesting standard of tea shoots for making the tea of commerce, 2 + bud, or 3 + bud which is based on the content of poyphenols, minerals, amino acids and saccharides The constituents level varies with soil, climate and management practices.

9.2 Type of teas

Green Tea: Tea leaf is processed with no withering and fermentation.
Black Tea: Tea leaf is processed with adequate fermentation/oxidation.
Oolong tea: Tea leaf is processed with partial fermentation.
White tea: Bud is collected and shade dried to produce needle teas.

9.3 Cell constituents

Black tea manufacturing technology involves disruption of the cellular integrity of tea shoots. This facilitates the mixing of substrates, polyphenol, and enzyme polyphenol oxidase present in the leaf. This results in the

Scientific Perspectives of Tea Plant Horticulture and Productivity
ISBN: 978-0-12-823444-0
https://doi.org/10.1016/B978-0-12-823444-0.00009-1

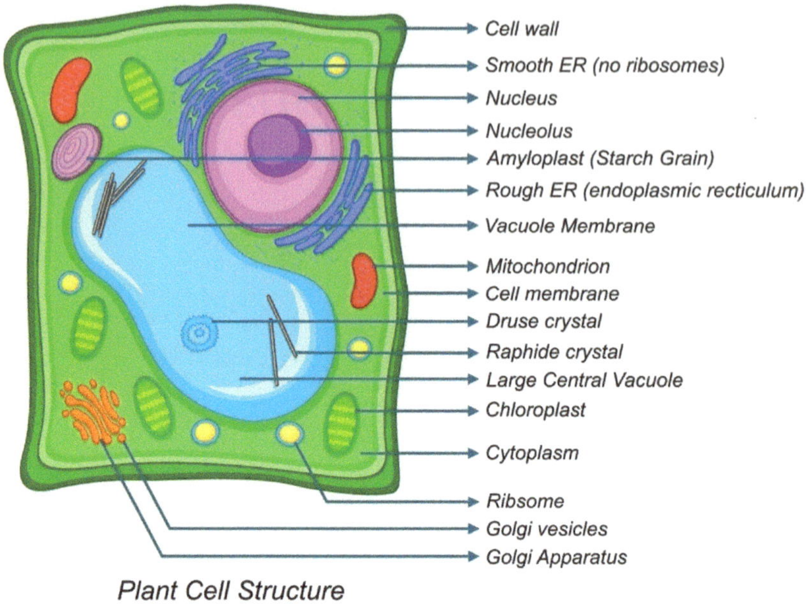

Figure 9.1 A cell with organelles and constituents.

initiation of series of biochemical reactions. Tea leaves are constituted by many cells which serve as manufacturing factories and store house of metabolic intermediates for the manufacture of tea (Fig. 9.1).

Vital constituents like polyphenols-flavonoids/catechins are compartmentalized in vacuoles in the intact cells while the cell sap in cytoplasm contains the organelles such as mitochondria, chloroplasts, and the nucleolus. In order to initiate oxidative processes the tonoplast of vacuoles should be ruptured, the constituents mix with the cytoplasm, triggering the enzymatic (PPO) reactions. The chloroplasts are the site of photosynthesis (RUDP-Carboxylase) producing the sugar molecules, serving as the building block/template for other metabolites. This is achieved by the shredding, rolling/twisting and the curl twist crush (CTC).

9.4 Black tea processing

(1) Withering partial removal of moisture; (2) rolling/disruption-size reduction; (3) fermentation: biochemical reaction in the presence of oxygen;

(4) drying: completion of moisture removal; and (5) sorting: fiber removal and grading based on size of the particles.

1. **Withering** is carried out for removal of partial moisture from the leaves (known as physical wither) attaining the optimum of 65%–70% moisture in the leaves and initiate preliminary biochemical reactions and changes (known as chemical wither). As the chloroplasts are still intact in withering stage, provision of fluorescent lights enhances the photosynthesis produced impacting on the quality of tea made.
2. **Rolling-Distortion-leaf conditioning-Shredding-CTC processing**: Preconditioning of the leaf is done with a device called a leaf shredder.
3. **CTC processing**: Shredded dhool is processed in Rotorvane as well as CTC rollers to obtain crushed, teared, and curled dhool for CTC teas. Withered leaves are passed through rollers directly to get twisted leaves for orthodox teas.

Fermentation–Oxidation:

Fermentation is a misnomer for the process of oxidation (removal of electrons) where oxygen is taken in releasing H_2O, triggering various enzymatic reactions, and releasing enormous heat. In order to sustain the temperature-dependent enzymatic reactions, humidity is maintained in the fermentation rooms/drum through humidifiers. Fermentation commences right from cells are brushed and extends until the enzymes are deactivated in the drier.

The principal chemical reactions during oxidation are depicted below schematically:

The primary reactions taking place during tea fermentation involves the oxidation of tea catechins into respective orthoquinones. Condensation of oxidized tea flavanol (orthoqinones) produces Theaflavin and thearubigin that are responsible for cup characters.

Catechins (EC-ECG-EGCG) Orthoquinone Theaflavin + Thearubigin

Issues to be addressed in tea plantations for quality crop:

In South Indian plantations, the plucking rounds are extended very often due to shortage of working force in the peak monsoon seasons. Absenteeism during festivals and paddy planting seasons in North east also adds up to this problem. Unless sorted out in the field, shear and mechanical harvest also contributes to this malady. As a consequence, coarse leaves and shoots are brought to the factory for manufacture, leading to grassy flavor (pacha taint).

However, research findings (UPASI and CFTRI) indicated that pacha taint could be minimized by introduction of vacuum packing and enzyme at one lit/ton of the dhool, while processing at second roller of CTC stage/ Rotorvane stage (if temperature is less than 35°C) with advantage on cuppage and appearance of made tea. The cellulase and pectinase enzymes solubilize the cell wall and middle lamella enhancing the total soluble solids (TSS), even from the coarse shoots and leaves in turn, the productivity and quality. Thus the use of pectinase enzyme at the maceration stage in the factory enhances quality of tea made especially during monsoon rush season as well as mechanical harvest with unavoidable coarse leaf and black leaf. As the enzymes are of natural origin-microbes/fungi, pectinase enzyme is permitted to use as processing aid during manufacture by FSSAI.

Drying: Drying of fermented dhool is to (1) terminate the biochemical reaction by regulating the inlet and outlet drier temperature; (2) to reduce the moisture to 3%–4% and produce tea with good keeping qualities.

Hence, the chemical reactions are terminated by firing in dryers, by regulating the inlet and outlet temperatures, forcing the moisture content to 3%–4%.

4. Sorting, Grading, Storing, and Packaging:

These are the final steps of processing/manufacturing of tea in the factories under hygienic conditions. Sorting basically converts the bulk teas into finished products confirming to trade requirements.

Pic. of factory, withering, fermenting, drying, packing.

Photos on

(1) withering;
(2) rotorvane;
(3) rolling;
(4) fermenting;
(5) drying; and
(6) sorting and packing;

Innovative upgradation required in processing and manufacture of tea (for the best quality and cleaner produce):

The tea processing factories are almost a century old and no conspicuous innovation in the design or the machineries has been taken place. The space utilized, machineries used, energy consumed, manpower involved need to go thorough up gradation, modification and refinement in the interest of cleanliness, labor saving, economy, etc.

The process of manufacture and packing has to be completely revamped in the light of the present day technological advancement and resources available. Although certain steps are taken up for continuous withering machine (CCW), and results are coming up, constant effort is required by the tea research institutes and IITs, and sped up to derive full advantage (Tables 9.1 and 9.2).

Table 9.1 Product diversification.

Value added teas	New product
Organic teas	Beauty-healthy-longevity-wellness teas
Black tea	Diabetic teas
Green tea	Decaffeinated teas
Oolong tea	RTD beverage -green and black tea
Orthodox special tea	Tea Concentrate -green and black teas
White teas	Tea wine
Purple teas	Tea vodka
Silver tips	Green tea extract
Golden tips	Theanine
Frost tea	Theaflavin
Flavored teas	Catechins

Table 9.2 Herbal value addition of tea.

Cultivar	Herbs	Part used	Botanic name
Tv-9, TV-23, P-126 & Av-2	Moringa	Leaves	Moringa sp
	Amla	Fruit pulp	Phyllanthus
	Ginger	Rhizome tuber	Gingeber
	Gloriosa	Tubers/seeds	Gloriosa sp
	Stevia	Leaves	Stevia sp
	Thulsi	Leaves	Oscimum.sanctorum
	Kilanelli	Leaves,fruits	Euphorbia sp
	Modakothan	Leaves/fruits	Cladiocarpus sp
	NilaVembu	Leaves	Androgynum sp
	Hibiscus	Flowers	Hibidcus sp
	Aswagantha	Roots	Withania sp
	Avarai	Flowers	Cassia auriculata
	Tinnaveli Senna	Leaves	Cassia.anguistifolia
	Adathoda	Leaves	Adathoda sp
	Turmeric	Tuber powder	Curma sp
	Aloe	Flesh, latex	Aloe vera
	Cardamum	Capsule seeds	Elataria sp
	Jasmine	Flower	Jasmine sp
	Nutmeg	Mace	Myristica sp
	Cinnamon	Leaves/bark	Cinnamon sp
	Jamun	Leaves/pulp	Eugenia sp
	Vettiver	Roots oil	V.zizavnoids
	Citronella	Leaves-oil	Citronella sp
	Centella	Leaves powder	C.asiatica
	Gymnema	Lvs-low calorie	Gymnema sp

9.5 Innovative processing/manufacture

Innovative reforms, upgradation, automation, labor, space, energy saving are some of the priority areas where R&D required, in the interest of the industry and the nation at large. Although some beginnings have been made in collaboration with technology institutions, their impact remains to be seen.

As new types of tea from high quality areas are emerging from many hill stations, with pruning, harvesting and processing, value addition of black and green tea to suit the taste and increased demand from consumers all over the globe, this area deserves attention and consideration of all concerned professional/management (Tea journey.pub/tea's immunity boosting properties).

The innovations are envisaged in the following lines meeting the demands from the emerging technology and younger generations.

1. Based on inherent high grown, premium quality, specialty tea of Assam, chinary, and hybrids—black tea orthodox, CTC, green tea, value added, and alone can be categorized, and GI allotted with standards for organic as well as conventional. Minimum expected ingredients, concentration of polyphenol antioxidants, theaflavin, catechins, theanine, and MRL, etc. should be enforced and implemented. Source of raw material, product of Indian tea, should be highlighted for traceability as safeguards.
2. The earmarked regions for the GI include the following:
   a. Assam Arunachal and pure Assam with chunky pubescent tips premium quality for export- premium black tea orthodox and CTC.
   b. High elevation chinary of Darjeeling, Kangra, Kulu valley Palampur, HP, the Nilgiris, premium orthodox flavor tea
   c. Himalayan river valleys with Assam, China, Cambod hybrids high inputs responding seeds

   TV 23 Assam, Cambod high potential clone suitable for GI region III for black and green tea.

   An young tea field in Kacharigaon, North Bank, Assam. TeenAli 17, a Cambod China hybrid clone selected by pioneers a hardy clone with dark maintenance leaves with light green yellowish shoots/tips, Responds well for high inputs and stress tolerant. *A. odoratissima* shade enhances the productivity with quality Temporary shade like Gora Neem and *Boga medulla* as cover crops, Citronella grass, Justicia, Adathoda, *Lawsoia alba*, and Lantana in the border as hedge filters the dust from highways/roads preventing the incidence of red spider mites. Intercrop with avocado is planned but postponed. Suitable for both black and green tea (Kacharigaon visit Report, 2009).

Teen Ali 17 is highly suitable for black orthodox and green and black tea. A potent clone for improvement.

TRF 4 selected by UPASI/TRF (pic below) is a versatile quality clone with stress tolerance, selected from the biclonal Seed Stock-BSS1.A, light-leafed China hybrid with flavor and quality suitable for mid to high elevations worthy of planting in southern India, Darjeeling, Himachal Pradesh, and similar climates and terrains and soil conditions, globally, such as Nigeria, Central Africa, Argentina, and so forth.

Kenyan China hybrid of Unilever Brothers, grown in Mambilla, MEISAMARI, Nigeria, Suitable for Nigeria, Central, Africa high Ranges for Quality orthodox black tea. The Mambilla, Meisamari region is a potential area for quality tea with diversification and value addition established by Unilever Brothers Nigeria, Limited, plc., established in 2003, a potent area for tea Expansion with 2000 ha, hilly terrain, planted, and established about 400 ha with unilever Chinary clone from Kenya, visited and submitted a positive report for revival (Report dated, Feb. 19, 2013 to the government of Nigeria, through Dangote Ltd.,Lagos, Nigeria).

d. Midelevations of Southern India with Assam-China hybrids with unique topography, terrain, field management practices such as young tea, shade management, pruning, harvesting, manuring practices including manufacturing/processing practices, and standard quality black tea with uneconomic price realization. Refinement and fine tuning of field management and processing for quality cleaner product are top priority and urgent for the survival of the tea industry (TRF4 quality productive clone).

TRF4 is a good quality clone selected from BSS 1(UPASI 8 x CR 60017—— TRF4).

e. a potent quality clone with productivity and stress tolerance, remains to be exploited for the hill plantations. Ideal for quality orthodox and green tea.

3. Innovative measures envisaged in energy, transport/conveyance, automation, Manpower saving, for implementation both in fields and processing factories.

Energy savers in tea processing-rollers for orthodox

a. Redesign the factory layout, bearing in mind the constraints involved at present Expert opinion and suggestions could be sought from individuals/organizations, the best one can be rewarded. The factory should be compact with vertical growth about three tier, adequate conveyers, thus area space is saved,could very well be utilized for Rain water conservation required for factory and field spraying operations with Advantage.

b. Solarization is one of the options for energy saving without interruptions

c. Transportation of harvested leaves from field to factory fast with minimum down time. The vehicle/carriers should be air conditioned for temperatures around 10°C and florescent lights, allowing the leaves active time in photosynthesis. The weighing and transporting should be handled quick and careful with least injury/stress.

d. Spreading the leaves in the driers, fermenting dhool in yards with humidification can be mechanized with monitoring of parameters with advantage, fixing the required gadgets and saving the labor involved.

e. The movement of withered leaves, dhool spreading for fermenting, drying tea, sorting, filling, and packaging yard all can be thoughtfully mechanized. Top downward as seen in hill stations.

Energy saving in tea processing-withering.

Unforeseen emergencies do happen when harvesting; processing times are hindered, despite the crop being ready for harvest. The damage to crop and quality can be minimized by machine harvest, using tractor mounted containers with light, ventilation, cooling to 10°C, and stored in similar capacity cooler rooms, along with withering floors; even for up to three days the leaves can be stored and manufactured. Thus overgrowth in fields by timely plucking, storage, and manufacture can be achieved with advantage. Adding up a leaf storage room with capacity provided with light, ventilation, and cooling (10°C), will ensure minimum damage. If the shoots are over grown, mechanical harvest is resorted and stored up to three days and manufacture.

To minimize the fiber content and grassy flavor of tea, ecofriendly proven product like Pectinaze Enzymes (Biocon, Bangalore) can be used in CTC second cut or pulverizer with vacuum filling.

Plate. Typical withering trough for quality clone P126 for orthodox black tea to be inserted.

Energy saving in tea processing—fermenting and drying

f. This is an energy-intensive area concerned with space, temperature, humidity, light, and oxygen involving the conversion of polyphenol, catechins in to Theafavin, Thearubigin, Theanine content etc contributing the ultimate quality of tea consumers prefer balanced and optimized. A very critical area of tea manufacture with personal attention of technical manufacturing assistant and manager, requiring large space and a costly drying, firing machines, with monitoring gadgets, needing expertise innovative technologies for economy in energy, space, atmosphere/environment for consistent quality, necessitating top priority of innovation and implementation with proven success.

Myura Tea Factory of Murugappa group in Valparai, Coimbatore designed along the gradient, withering till sorting and packaging, automated through conveyer belts, monitored with Instruments, saving space, work force/mandays, producing about 20 L kg/year/annum with consistent quality under very congenial environment and resources. A typical processing factory for hill country of Southern India, saving energy, space, mandays,with consistent quality/price realization, pooling the technology with sincere dedicated management and work force, an egalitarian society with harmony/peace envisaged.

9.6 Diversification value addition and marketing

Tea Industry organized and opened as monocrop, invariably rainfed, regulating the ambient environment with shade, cover crops, mulching, building up the bush Architecture, horticultural in evolvement of the corporate, pioneers, entrepreneurs specialized as Industry. In course of time small scale growers of tea emerged by local entrepreneurs with Govt supports creating employment opportunities for the unemployed skilled enterprising local population. This created a healthy competition with strict Rule Based dominant Inspectorate, corporate and entrepreneurs. Another important plus point of small tea growers is their integrated approach with diversification and value addition and freedom of marketing with less overhead charges, thereby competitive cost of production.

The indigenous vegetation of Horticulture, spices, herbs came handy for diversification and valuation meeting the demand and supply with advantage. Thus the STG are the pioneers of specialty teas, like golden tips, silver tips, white, purple tea, value added tea with, ginger, lemon, mandarin in Assam, Arunachal, adopting ecofriendly GAP with due concern for environment and consumers. Similarly, the small tea growers of Nilgiris are successful in organic hill produce of tea like Frost tea, orthodox, Green tea. The Sri Lankan Dilmah tea also pioneers in value added Jasmine tea etc with innovations. The KTDA of Kenya are also pioneers evolving package of practices, blending, marketing their tea competitively on cooperative basis. The TaTa Amalgamated Tea of Assam (Hattikuli) also are pioneers inter-cropping black pepper as model demonstrating the feasibility of growing multicrop/diversification, and value addition, marketing overseas as Tetley Tea, USA.

Consequent to the pandemic situation, imparting immunity through herbs including tea has been recognized globally. Thus, it is a splendid timely opportunity for tea plantations of India, Sri Lanka, and Central Africa to take full advantage of the situation to diversify, with potent proven horticultural crops, spices, and herbs. The herbs can be added with tea blending with the economic plant parts like leaves, flowers, or processed fruit, flowers, as the case may be. Alternatively, the active ingredients (Alkaloids) can be isolated and used for value addition. In the interest of the well-being of consumers and producers, required law/Plantation Act can be amended to facilitate the introduction, diversification, processing, value addition, packaging, patenting, marketing, and so forth.

Prospective crops and the alkaloids for introduction as intercrops in tea plantations are given below:

Plant name parts used Alkaloids/Steroids/ai
Avocado/Macadamia nuts pulp/seeds Vitamin B17
Mandarin/lemon Fruits: rind/skin Vitamin C
Sour Sop/Graviola Fruits/Leaves Proven anticancer
Turmeric, Ginger Tubers Curcumin, Gingiber
Papaya Fruit latex Papain
Neem Leaves Azadirachtin
Moringa Fruits/leaves Vitamins, Minerals
Black pepper Seeds Pipercin
Cinnamon Bark, leaves Alkaloids
Cardamom/cloves Capsules, seeds Alkaloids for dental care
Nutmeg Mace, seeds Aphrodisiac
Andrographis-NilaVembu Herbal Leaves Bitter alkaloids, steroids/ antivirus
Jasminum grandiflorum flower buds flavor, antiseptic, aesthetics
VanillaPlanifolia Cured beans Vanillin, Beverages
Mulberry Fruits Vitamins, Minerals
Mudakathan herb Cardiospermum halicabum Rheumatism/Arthritis cure

The option for diversification and value addition are open to the corporate as well as small tea growers of all regions of both North and South India, provided they abide by the GI quality standards, adopting integrated, multiple cropping with ecofriendly GAP, avoiding hard prohibited chemicals and herbicides, meeting the MRL. The upcoming small growers are well placed for this innovative project, since they are following Integrated Agricultural Practices, established most of the suitable crops in their gardens,most of them are familiar with traditional indigenous medicines, fruits, spices, herbs. They can convert part of their existing garden properties and initiate the program with out time lapse. India is the nation blessed with various terrains, climate, biodynamic, crops like fruits, spices, herbs well utilized for the Indian/Indigenous medicines and herbs, spices for homeopathy, Ayurvedic, Siddha traditionally, which are now recognized as effective, harmless, against some of the pernicious virus, bacterial diseases emerged as the result of over exploitation of nature, industrialization, natural hazards, and pollution of water resources, eroding the nature.

The present pandemic global situation is an eye opener for revival of ecofriendly alternative medical care. India has the rich potential equipped with technology, know how, resources like tea coffee, rubber plantations, spices, horticultural crops, and herbs which can be utilized with advantage by both small, big growers of tea plantations to revive and boost the growth of tea plantations/economy.

The Government of India, under the leadership of the Prime Minister Narendra Modi is keen to help the growers launching various innovative horticultural, food preservation, soil-water conservation ecofriendly, diversification, and value addition projects, implementing the work force welfare measures, aiming the congenial, egoist friendly relationship with harmony of the garden owners, corporate, dependent work force, assuring the bright future for all the stake holders of the tea industry including the local population, creating employment opportunities. Any impediments/blockages implementing these proposed Tea, Horticultural, North East oriented proposed innovative, reforms, upgradation, technology development, soil and water conservation including flood control measures are receiving the attention and ready to bring in amendments of Plantation and labor acts to facilitate the quick implementation of these vital valuable positive projects:

1. Revival of the Tea industry with GAP, diversification, value addition
2. Horticulture dev. and food preservation, welfare measures for work force
3. Flood control of Brahmaputra River restoring transportation and marketing in the interest of Socialistic welfare of the people and nation's economy.

Considering the herbal wealth, biodiversity, technology development, proactive government/leadership, and pandemic global situation, Assam, being the leader of North East, the paradise on Earth, can take the initiative and lead for reforms, soil and water conservation, introducing ecofriendly GAP for tea and horticultural crops including preservation, processing, and aggressive marketing bringing prosperity through creation of employment opportunities, infrastructure development, on priority leading the way for other tea regions of the country and worlds at large. Let us all hope and pray for the best.

CHAPTER TEN

Pollution of water, air, and toxic chemical elements

This is one of the serious problems we anticipate in plantations due to industrial revolution in the modern world. In places like Bengal the soils are high in Arsenic and Chromium. Unless the level reduced through leaching or amendments, it may show up in plantations. The population will be affected with health hazards.

Higher, toxic levels of any particular or combinations of heavy metals like Pb, Cr, Hg, As, Ag, and so forth will affect the population of microbes, fungi, and beneficial predators and parasites, including mycorrhiza, which will affect the health and productivity of the bushes and human population. Flooding and leaching the soil and incorporation of heavy dose of organic matter, the toxic elements level can be reduced to some extent.

Brick kilns, oil refineries, coal/dolomitic lime quarries and mines fertilizer/chemical, cement factories, and so forth should be located at a safer distance so that the plantations are not affected by the air, water, and soil pollution. Any polluted atmosphere impacts on the photosynthesis and metabolic functions of the plants causing irreparable damage affecting the health, productivity of the plants and quality of the produce. This is another important area for research and development (R&D).

In addition to the natural pollution, the agro-inputs used in plantations, and farm fields also contribute especially, when intensive monoculture like plantations, due to continuous excessive heavy doses used that too from unauthorized unrecognized sources. The present global situations in tea plantations experiencing yield stagnation, with high cost of production coupled with poor value realization is a pointer that we have over exploited the soil and environment paying a price.

Unless we correct the situation amending the damages to the soil, environment, and water sources, rationalizing the inputs and appropriate restorative measures, the future is bleak/uncertain. We have to revert back to ecofriendly good agricultural practices (GAP), simultaneously withdrawing the pernicious chemicals and herbicides from plantations. Encouraging/strengthening the vegetation buildup. Role of plantation managements on pollution control, providing cleaner air, water, and clean environment

Scientific Perspectives of Tea Plant Horticulture and Productivity
ISBN: 978-0-12-823444-0
https://doi.org/10.1016/B978-0-12-823444-0.00010-8

protecting the biodiversity ensuring the health and wealth of the people and nation at large.

Restorative measures, suggested steps with time frame for buildup of cleaner environment in plantation areas on priority:

1. Utmost care and attention in afforestation, shade management, strengthening wind barrier in all plantations tea, coffee, and rubber including spices ensuring ecofriendly natural farming/GAP avoiding all hard chemicals.
2. Prioritize intercropping, multicrop system integrating with dairying, diversification, herbal indigenous medicines (value addition), and so forth, preserving/optimizing the biospheres of soil/plant environment for sustainable productivity with quality, rationalizing the bioinputs, standardizing the package of practices impacting on the reasonable/affordable cost of production (COP) and price realization.
3. Rain water harvest, storage, flood control in Brahmaputra, desalting, cleaning, deepening rivulets, draining the excess water from the gardens in to the big river bringing down the water level of the gardens below root zone maintaining the river water level at least three meters below the bund. Forage crops leguminous shrubs and trees on the banks/shores of rivulets, canals preventing soil erosion. Blockages if any in the rivulets should be periodically removed/cleared and outfall ensured reducing the pressure on the garden drains. Flow restored by coordinating efforts of CPWD with gardens on sharing basis.
4. Ensure rainwater harvesting in plantations, conserving water, soil, and air with minimized level of pollutants, protecting the biodiversity and environment enhancing the quality of living in plantations.
5. Ensure the flood control, drainage, water level below the root zone ensuring precision/natural farming with zero budget maintaining the soil moisture:air ratio at 60:40 facilitating the enhanced metabolic functions of the plant ensuring the productivity with quality and longevity/life span of the plants.
6. Nutrients management is important for sustainable productivity with quality. nevertheless, optimization of all the desirable/vital soil parameters such as soil, pH, humus content, microbial population, root proliferation, healthy roots, and canopy are important to effectively absorb, utilize, and assimilate nutrients applied and carbohydrate produced for cost effectiveness. Thus, soil reclamation plays an important role on health and productivity of tea plantations for sustainable productivity with quality, rather than the quantum of nutrients and time/

form applied As these NPK fertilizers are subsidized, the growers have a tendency to be liberal and overshoot the limits. Though the replacement theory, which is a guideline for fertilizer schedules, caution is required to fix a ceiling to minimize the damage.

7. The Om and fertilizers mixed 50:50 with biocides applied in split doses, cheeled/hoed, and incorporated.
8. Winter operations like drain cleaning addition of soil amendments for soil Reclamation optimizing the soil pH/acidity, hoeing, and aerating the soil with humic acid will multiply the microbial population including VAM enhancing the feeder root proliferation, early ground cover, and P utilization enhancing the absorbency of nutrients without nitrite buildup thereby the soil acidity inducing micronutrients deficiencies that could affect the productivity and quality.
9. The fertilizers when spread on the fields are exposed to bright sunshine aggravating the volatile loss of ammoniacal nitrogen up to 10%.
10. In steep terrain slopes like southern India and Sri Lanka up to 40% nutrients applied are soil washed and leached out contaminating the soil, streams, rivulets, reservoirs etc affecting the salt concentration osmoticum of water uncomfortable for aquatic planktons, algae, fish, other amphibious reptiles tortoise vanish in the course of time harming the useful versatile biodiversity of hill stations and water resources. Thus it can be safely inferred from the past 100 years of experience in tea plantations, excessive higher doses of NPK fertilizers without balancing with organic matter/humic acid based on the replacement theory without ceiling incalculably damaged the soil, plant health affecting the productivity, quality, and longevity of the bushes. The washed-out leached macro/micronutrients contaminated the low-lying streams, rice jute fields influencing spectrum of types of fish, thereby the livelihood of farmers, fishermen by none of their follies but the over ambitious unscrupulous overzealous planters. Nevertheless the time has come, the mistake fault has been realized of blindly adopting the replacement theory to 50% in dose, in nutrient management, and time has come to correct/rectify the damage, reclaim the soil with abundance of organic matter reverting back to ecofriendly GAP following the microbial, micronized nanonutrients, and biostimulants through evolved package of practices retrieving the plantations from chemical adversaries and move forward to value-added organic products of Assam, Darjeeling, and the Nilgiris, encashing improved quality enforced under the GIs of the tea board of India. For ensuring

boon and prosperity of Indian tea, minimizing the impact of increasing chemical pollution world over quickly saving the consumers, especially children and senior citizen producers.

Awareness of chemical contamination and pollution on the health and productivity of tea plantations of North India, inclusive of the workers and depending on the rural population have come, and GAP with value addition and diversification has been introduced both by corporate as well as STG. It is just a matter of time that the Assam tea is declared as organic, taking all the benefits accrue while overcoming the ill effects/damages due to the contamination and pollutants in the rivers, soil, and produce. It will be a win-win situation for all the stake-holders of the tea plantations in North India. However, the situation is different in southern India, where the mind-set of planters have to change to redeem themselves from the chemical-oriented to clean cultivation preferring quick knock down hard chemicals for plant protection and the misplaced high inputs of chemical fertilizers on the basis of crop replacement theory. Transformation of refined proactive innovative field and processing package of practices biased on integrated nutrient management, bioproducts, need-based use of micronized nanomicronutrients with biostimulants, optimizing the soil and plant biospheres with rationalized inputs, need-based foliar application, shade management, and so forth will give ample opportunity to overcome the hazards of chemical pollutants damaging the water and soil resources, safeguarding the livelihood of millions of plantation workers and depending population. Nature should be allowed to correct the damaged situation in plantations. The plantations will recover and flourish, helping the entrenched worker population restoring the cleaner environment and biodiversity. This will satisfy the government protecting the environment laws and forestry, a win-win situation expected no sooner than expected. Nevertheless a strong reorganized R&D under the national experts committee of science and technology under the PMO should be formed to evolve and streamline the improved alternatives with a package of practices evolved, enforced, and monitored periodically for successful implementation of the national, agriculture, horticultural, and plantations acts at the international level with coordination and cooperation of the Food and Drug Administration/FAO successfully.

It has been confirmed that soils with optimum pH, OM shade, drainage, and microflora respond abundantly well up to 7000 kg MTH at low level of 100 kg N/ha balanced with Nk at 2:1 ratio broadcast in split doses and incorporated by hoeing. For effective utilization of applied nutrients timely effective harvest of shoots is essential, manufactured, and marketed aggressively, taking advantage of the demand and supply, and the prevailing market situations. Taking advantage of the value addition is worthy of mentioning to take advantage of added value premium prices realized. Because of the continued excess dose of fertilizers without balancing organic manures most of the water resources of Southern India and Sri Lanka have been contaminated by chemical salts, making the water resources not safe for drinking. Thus the higher doses of N spoiled the water and soil with soil wash and leachates. The soil and plants are exhausted and weaker, exhibiting micronutrient deficiencies predisposing to resurgence of virulent pests and diseases making more vulnerable. Most of the beneficial predators and parasites have been wiped out compelling the planters only quick knock down chemical-oriented with escalating COP concomitant with all the hazards of contaminated soil and water resources.

The concerned forest, environment authorities, and tea plantations (UPASI) are aware of the present situation and seriousness of the problem. We planters have to amend our activities in hill stations, reform the package of practices minimizing chemical load in plantations safeguarding the soil, water resources restoring the glory of our hill stations as a flourished tourism spots, and live and let live with nature peacefully.

If the planters are reluctant for reform/innovative changes, the government has no other alternatives but to declare the lease period is over and expired, revoking the long lease, with a takeover of the plantations restoring the natural sholas for the benefit and enjoyment of the nation and people with health and prosperity.

Thus, it is time to streamline these soil and plant biospheres, reduce the chemical load rationalizing the inputs through integrated nutrient management, thereby the COP, doubling the productivity while minimizing the contamination of soil, water resources protecting the health, and welfare of the workforce and depending population on tea plantations all over the world through adopting ecofriendly GAP.

Albizzia S1 shade Tree

CHAPTER ELEVEN

Current problems and remedial measures required in tea plantations

Contents

11.1 Present situations

The plantations have completed almost a century of exploiting the natural resources. Nevertheless the growing conditions, namely, climate, soil, and bush health, are not the same due to over exploitation of the resources. The bushes are weak and crop is stagnating. The weak bushes are more vulnerable to stresses, pests, and diseases.

The climate change, indiscriminate and continuous use of pesticides, fungicides, herbicides, higher dose of NK causing hyper acidity contributing hard pan, vacancies in both tea and shade altering the microclimate favoring higher incidence of pests and diseases. The tea bushes and shade trees are under chronic stresses due to exhaustion, deficient in starch reserve, micronutrients deficiencies aggravated by the toxicity of nitrite build up, hardpan, depletion of humic acid level in the soil, and starving the beneficial microbes population. Besides the soil biosphere, plant metabolic processes are affected by the buildup of glyphosate in the system accelerating the oxidation of metabolites, causing the deficiency of micro and secondary nutrients like Zn, Mn, Mo, B, and S, a typical stress symptom of higher Maximum Residue Limit (MRL) of the systemic herbicide glyphosate.

Thus the present situation of damaged soil, plant biosphere, and biodiversity, inducing chronic stresses, affect the productivity and quality of tea plantations, facing stagnation, poor value realization, yellowing of leaves, defoliation and dieback of bushes, and trees causing vacancies up to 30% in most of the affected plantations.This is a serious situation warranting the priority attention of the planters and governments.

Scientific Perspectives of Tea Plant Horticulture and Productivity
ISBN: 978-0-12-823444-0
https://doi.org/10.1016/B978-0-12-823444-0.00011-X

11.2 Major problems faced by tea plantations

The following are the major problems faced by most of the plantations making the operations unremunerative due to escalating costs of production:

1. Low pH (high acidity) of the soil
2. Low organic matter content resulting in depletion of microbial population
3. Incidence of root diseases and frame canker effecting mortality of bushes, and thereby vacancies
4. Occurrence and resurgence of new types of pests both in tea and shade trees pushing the plant protection cost high impacting on the cost benefit ratio and value realization
5. The climatic changes brought in by natural calamities and deforestation
6. Exposure of the bushes and shade trees to adverse, climate, soil, stresses (biotic and abiotic), overexploitation of the potentials of the bushes without compensation, rectification, or amendments, which led to vacancies
7. The escalating COP, with low price realization due to excessive use of inputs not only affected the productivity but also the quality and price realization due to chemical residues, unclean tea
8. Consequent to the failure and withdrawal of many corporate, welcome changes, and transformations, amalgamation, consolidation, and regrouping as well as a change of hands with financial and technical backing took place, which is a sign of a bright future

Ideal conditions and measures are suggested to enforce in the fields a double hedge planting structure with a population around 15,000/ha in contour.

1. Staggered
   a. Soil with adequate OM, with a pH range between 4.5 and 5.5, soil moisture:air being 60:40
   b. Adequate drainage without water logging or moisture stress
   c. Optimum (leguminous) shade with regulation during monsoon, aeration, and humidity
   d. Soft herbaceous weeds preserved to serve as mulch, reducing the moisture loss through evaporation, encouraging microbial population and feeder root proliferation
   e. Meticulous cold weather operations-cleaning and deepening the drains, weeding, mulching, and water conservation measures

f. Pruning the bushes in winter after resting at least a month to enrich the starch reserve
g. Add a new leaf on top of the plucking table (canopy) by October (Puja holidays) and commence stress amelioration measures (foliar sprays) in winter
h. Judicious foliar spray of micronized micronutrients with amino acids: bioproducts help photosynthesis and growth, increasing quality and productivity
i. Harvest the crop periodically at 8–10 days interval by standard plucking, ensuring 32 rounds/year
j. NPK manures at 2:1:2 ratio, inorganic:organic at 2:1 ratio, applied in 3–5 splits
k. Commence with prophylactic measures with herbal and bioproducts alone in combinations followed by soft proven chemicals encouraging the predators (natural enemies). Any spray should be repeated within a fortnight to take care of the escapees or new generations
l. When build up has taken place better to give the appropriate knock-down chemical and switch back to softer options, avoiding residue and quality problems

Future of Hill Tea plantation

CHAPTER TWELVE

Priority areas of research for the preeminent position of Indian tea plantations

Contents

1. Plant improvement through breeding is a continuous process to produce elite cultivars meeting the advanced cultural requirement of the industry
2. Cell culture, micropropagation, standard media, and package of practices should be worked out as nursery media and space will become scarce

Scientific Perspectives of Tea Plant Horticulture and Productivity
ISBN: 978-0-12-823444-0
https://doi.org/10.1016/B978-0-12-823444-0.00012-1

3. Evolution of ecofriendly package of practices for cleaner product of India with sustainable productivity
4. Research on the impact of soil amendments on health and productivity of tea with cost economics
5. Integration of horticultural crops and dairying with tea for sustainable healthy plantations and population
6. Intensive research with quick results on usage of microbial and herbal products, alone and in combinations (consortium) for plant protection and management practices, evaluated and proven in regional trials that are cost-effective and adaptable for the entire region for a green northeast
7. Introduction and evaluation of different popular indigenous leguminous trees as shade as well as agroforestry (wind barrier) to screen and select the most suitable trees for our plantations. In the light of the climate change and ecofriendly management practices, additional requirement of biomass-compost, oil cakes, and fuel needs to be met indigenously in the coming years; follow-up on agro-forestry and dairying required
8. Diversification and value addition of tea, meeting the global demand and consumption, research on new tea cultivars, manufacturing/processing, herbs, development of package of practices for production, processing, blending, packaging, and improving the value realization need further exploration to maintain the tea industry's relevance.
9. The energy consumption, type of machinery used, layout of the factories, methodologies adopted for manufacture, sorting, blending, packaging, and so forth need to be modified and upgraded adopting the developed technology, for aesthetics and the economy.
10. Darjeeling still deserves an elite biclonal/polyclonal stock replacing Nanda Devi. Darjeeling, being the Champagne of India, is sought after globally; the productivity should be increased, retaining the unique flavor and distinct characteristics.
11. An integrated ecofriendly package of practices with suitable leguminous shade, cover crop, soil conditioners, elite clones and seeds, with indigenous inputs and package of practices, and aggressive marketing with diversification and value addition are very much essential on priority to catch up the lost logistics/strategies.
12. So far we have concentrated on the top-canopy biosphere exploiting the productivity with inputs management. Least attention has been

given to the soil biosphere, which plays an important role on plant health with sustainable productivity and quality. The time has come to devote enough attention to build-up and optimize the biosphere of soil to carry forward the productivity with quality sustainable. Research is required to standardize and optimize the soil biosphere parameters region for sustainable higher productivity, taking care of local terrain, climate, cultivar and productivity, maintaining the health of soil and plants, extending the life span of the bushes. Strengthening/restoration and maintenance of the soil biosphere for all perennial horticultural crops, are the top priority areas for the entire world, for food security and health.

12.1 Importance of the soil constituents, strengthening, and handling for sustainable productivity and cost-effective management

12.1.1 Principal constituents of soil biosphere

1. Humus and Humic acid
2. Beneficial Microbial Population
3. Root System- Feeder Roots, primary, secondary, tertiary roots
4. Colloids, aeration and preservation of micronutrients

1. **Humus and Humic acid**: The organic matter (OM) content of the soil plays an important role in generation and build-up of humus/humic acid. The microbes, which use the Organic matter as food digest convert into humic acid for their multiplication and build-up. Humic acid is a good buffer regulating the acidity/alkalinity tying up the concerned nutrient elements regulating/mediating the soil pH and availability of macro nutrients, protecting from leaching. Thus ensures the prolonged availability, continuous supply of essential elements for sustained metabolism. Humus ensures the crumbling colloidal structure of soil, contributing for the aeration of soil and sustaining many enzymatic reactions like nitrogenase, reductases, and ketoses, providing energy for uptake and metabolism.
2. **Beneficial Microbial Population**: The microbial population, directly contributing to enrich the soil, are many, and their population decides their contribution, which is in turn governed by the richness of the soil

with good aeration and nourishment. Acetobacters, phosphobacter, closterium, and mycorrhiza are some of the useful bacteria and fungi that directly contribute to generating a nutritious environment. Tricho-derma, Pseudomonas, Verticilium, Metarrhizium, and *Bacillus subtilis* are some of the beneficial microbes isolated from soil, multiplied, and used as biocides and biofertilizers. A good soil with optimized pH, OM, EC, aeration, and colloidal with adequate humus takes care of these beneficial microbial populations. Most of the soil-borne pathogens eelworms and harmful species are taken care by antagonistic effects of rich beneficial microbes build-up. In addition to these microbial population, soil accommodates the beneficial flora and fauna, which also contribute to building up the soil rich in nutrients, aeration supporting the plant health, productivity, and sustainability. The earthworms and mycorrhiza are well exploited for enriching the soil to sustain the productivity in an ecofriendly way.

3. **Root system: Feeder, Primary, Secondary, and Tertiaries**: The role of root system for anchorage and absorption of water along with nutrients is well known. The feeder roots spreading in the surface soil are important in absorbing the applied inputs like NPK mediated through enzymes like nitrite reductase for conversion into nitrates. The root exudates also play a favorable role influencing the soil media facilitating enzymatic reactions supplying the cofactors, vitamins sustaining the metabolic processes. The mycorrhiza connection with roots help phosphate mobilization symbiotically. The roots are the major storage reservoir of starch. The status of root reserve being the index of health plays a dynamic role on growth and development of the entire plant. The rhizosphere area is almost protected and guarded by the root secretion with wide ranging potentialities. The water table in the ground controls/determines the depth of root system. In the terrains of hills of Southern India and Sri Lanka the roots go deep up to 3–5 m exploiting the nutrients and water, whereas in North India plain plateau, the root system is shallower and restricted. However, this constraint is compensated by the rich, sandy, loamy soil, coupled with conducive favorable weather impacting robust vigorous growth and sustained by the silting/sedimentation, leaching the salts, and aeration through the humic acid content, enriching the soil periodically.
4. **Colloids, aeration, and preservation of micronutrients**: Under favorable soil, plant water relationships, organic matter (OM) content

of the Rhizosphere, the humic acids preserved and, leaching of micronutrients prevented. With adequate moisture and aeration, the soil forms crumbs in the colloidal form promoting the enzymic activities of metabolism utilizing the preserved micronutrients. These increased activities of the favorable soil sphere, in turn, stimulate/trigger the canopy metabolically coordinating, the functions of the growth and development of the pant mediated by the hormonal balance, source-sink ratio, partition of assimilates through enhanced photosynthetic assimilation impacting increased productivity. These are correlative plant functions that can be regulated through judicious handling of the field and input management. Thus a careful, judicious handling of the soil biosphere is the priority area for utilizing the full potential of the canopy biosphere, exploiting the prevailing favorable weather with rationalized inputs. It should be a holistic approach, balancing the potential with demand, adopting ecofriendly, cost-effective field and input management for sustainable productivity with quality. It will be a win-win situation, rewarding all stake holders.

12.2 Tail end crop and stress management practices for Assam tea: prophylactic and ameliorative measures

The tail end crop in Assam is precious scientifically, warranting special extra care due to the envisaged 2019 *El Nino* Impact I certain basic scientific facts

1. Plants perform better in ideal condtions than in stress
2. A weak plant is more susceptible/vulnerable to pests and diseases (biotic and abiotic stresses)
   a. Judicious prophylactic and palliative/ameliorative measures reduce the impact, minimize the losses, preserving the health, and productivity sustainable.
   b. Any measure proposed should be need based on the local producing conditions, cost-effective, affordable, and amenable for timely execution.
   c. The cost benefit ratio should be favorable and advantageous to the management and plantations.
   d. Nurseries/germplasm, sensitive valuable crops, and fields deserve priority in implementation

- **e.** Plan of action should made, after thread bare discussions kept ready with inputs and materials required for timely action, e.g., inputs, machineries, accessories, extra manpower, fuel, and so forth
- **f.** YT, HRP, RP sections/fields deserve consideration and priority due to their vulnerability
- **g.** The measures could be short, medium or long term up to six months in the interest of the management and property aiming sustainable productivity
- **h.** Science never fails, hard work and perseverance pay, everything is possible with positive attitude and dedication. Let us do our best and leave the rest to Nature. Nature cooperates with sincere efforts

12.2.1 Stress amelioration measures for the winter months November – March. a. Soil, b. plants, c. inputs & d. action

1. Take soil sample in October; analyze and get optimized by pH, OM, soil moisture, and aeration
2. Plant: Optimize the canopy thickness by leaf addition, balance between source/sink thru effective, timely harvest of the crop: labor, materials, and machineries organized
3. Soil health and preservation of biosphere of root zone and canopy play a role for response to inputs
4. Rationalize the inputs for ecofriendly integrated, cost-effective economic package of practices toward sustainable productivity with quality.

12.2.2 Details of practices: soil nutrition, aeration, biosphere

1. The final/last split of manures NK dose is reduced by 50%, matched with organic manures, compost/cattle manure/biomass, tank silt, oil cakes, soil amendments, broadcast and incorporated by hoeing, and including removal of overgrown weeds
2. Add a mature leaf on the plucking table of the canopy as a moderator, filling up the gaps/patches
3. Foliar apply fortnightly the cocktail of NK + nano micronutrients and amino acid (half strength)
4. Ensure effective, timely, standard plucking, and balancing the demand and supply (source-sink) channels of the plants for crop, quality, and sustainability

5. Give a prophylactic foliar spray cocktail of Trichoderma + Pseudomona S + Metarrhizium in continuation of micronutrient/*B. subtilis* spray.
6. YT, HRP, RP sections
   a. Postpone/withdraw application of fertilizers, herbicides spray, instead commence sprinkler fertigation fortnightly preferably in the evenings to mitigate the stresses
   b. Mulch the soil heavily with OM, thatch, intercrop with leguminous herbs, shrubs, for shade and biomass, preserving the bio-sphere,microclimate with build-up of valuable microbes
   c. Massive agroforestation/greenery encouraged in vacant areas in boundaries as barrier, shade trees and avenue trees
   d. The value added intercrops like black pepper, cardamom, spices, fruits, and vegetables deserve special care and attention

12.3 Packages for improving crop during quality seasons second/autumn flushes regions: Assam, Darjeeling, Nuwareliya, and the world's best black teas

Table 12.1 Crop Promotion and Quality Improvement for Black Tea quality areas.

Variables/ practices	Assam/Arunachal	Darjeeling	Nuwareliya
Soil/Water conservation	Acidic, alluvial sandy loam, river alley, plateau low-opt soil carbon excess water/ monsoon, def. winter	Acidic high OM, hilly slopy, high water loss nutrients	Uva similar to Darjeeling warm dry humic in winter
Climate, terrain	Bright sunshine cool humid weather monsoon/Autumn, cold dry winter, drainage, border hedges, grass embankment	Cool, humid,bright sunshine monsoonic, autumn, cold winter	A tourist center staggered trenches required

(*Continued*)

Table 12.1 Crop Promotion and Quality Improvement for Black Tea quality areas.—cont'd

Variables/ practices	Assam/Arunachal	Darjeeling	Nuwareliya
Shade status	Leguminous shades regulated diversification with spices, herbs fruit trees and dairying, value addition encouraged	No or poor shade!? Cover crops, low shade wind barriers avocado, mandarins, graviola, pineapple, ginger encouraged as intercrop	Boundary wind barriers, low shade, cover crops arrest soil erosion, nutrients evapotraspn conservation build-up microbes of
Weed types	Pernicius weeds hand pulled/cut	Grassy soft weeds preserved	Leguminou coverm, foliage crops encouraged
Cultivars	Typical Assam/cina hybrids	Typical chinary	Chinary cambod hybrinds
Pruning	After inter flush domancy	Monsoonic second flush	After interflush dormancy with rest
Cropping period	Focus on early tail end crop	First second autumn flush	Quality Uva season
Inputs	INM@50:50, 2; 1; 2,100NK/ha,3 splits, GAP, bionanonutrients, with soil amendments to ensure soil biosphere optimized/Reclaimed	Ensure soil, pant biospheres optimized by GAP biased on dairy bioinputs, local herbs with value addition	Ensure all inputs locally made available self-support and sustainable
Intercropping/ Diverse	Positive	+ve winter herbs fruits vegetables	Positive with nuts spices fruits, flowers herbs
Harvesting	Mechanization with foliar spray	Machine harvest with nourishment concern add value stevia herbs flowers	Diversification value addition ad effective marketing
Processing	Orthodox green tea specialty teas	Orthodox back tea	Diversify value add ortho
Marketing	Innovations required	Streamline	Regulate demand supply
Innovations	Pruning harvesting processing	Productivity	Economy

PS, Rest the bushes at least a fortnight before knifing, to build-up adequate starch reserve.
LP, HRP will require at least a month resting to get the reserve around 20%.

Check the reserve status with Iodine(Tincture Iodine 1 mL mixed/diluted with 9 mL clean water) test on the spot with pencil thickness roots. Allow 2–3 min. for color development (Table 12.1). If the color is brownish resting period extended. Starch reserve is an index of bush health, which builds up approximately at 4% per month resting.

12.4 Present situations/conditions of tea plantations, restoration measures, suggested with time frame

Growing conditions of tea in the past and present are compared, emphasizing need for redefining the field management practices, focusing research on priority areas.

Field situations in terms of soil, shade, water management, spectrum of pests and diseases, planting material that contribute for the productivity warranting upward revision of the nutrition input, its impact on productivity needs to be addressed. Considering the labor strength and productivity, mechanization of field operations is an urgent necessity.

The manufacturing conditions warrant radical changes in terms of factory layout and machineries used, focusing on energy conservation, producing a consistent, quality, and cleaner product while meeting the demand of the consumers.

Thus, the development and fine-tuning of field management and manufacturing practices including mechanization are the need of the hour for the sustainability and economic revival of the tea industry.

Present situation and conditions of tea plantations in India.

Consequent to the intensive Management Practices of high inputs NPK without balancing or ceiling the soil parameters, biosphere, and plant health were damaged irreversibly in the past 20 years, weakening both tea and shade trees, predisposed to resurgence of virulent pests and diseases, not amenable to routine, recommended available chemicals and dose. Synthetic pyrethroids, phosphatic, chlorotic chemicals, paraquat, glyphosate, and herbicides were used repeatedly in higher doses in their anxiety to control the situation. This resulted in exhaustion of bushes, going through chronic stresses resulting in stagnation of productivity with poor quality, value realization, and escalated COP, leading to bankrupt properties, including leading renowned corporate, due mainly to mismanagement and lack of cash flow. Aggravated further due to lack of adequate latest technology available to face the critical situation. Restoration and revival of Tocklai, TRI, Jorhat, Assam, PIN785008 INDIA with adequate infrastructure, expertise/

personnel with financial support, and administered by a Scientific Advisory Committee directly under the Prime Minister of India. It is a priority area for the early recovery and prosperity of the industry and the nation at large.

12.4.1 Restoration measures envisaged during 2020–22, North Indian tea

1. Soil amendments/reclamation
2. Enriched OM addition: Shade, cover crops, mulching, conserving soil moisture, root proliferation, and microbial build-up
3. Reduction and withdrawal of herbicides
4. Ecofriendly package of practices (GAP) to manage the pernicious pests and diseases

   Impact of:

1. Low pH of soil on health and productivity of tea bushes (adverse effect)
   a. Burns up the feeder roots system affecting the nutrient absorption
   b. Eats up the carbohydrate reserve weakening the bushes through accelerated respiration
   c. More banjification, poor crop due to bush weakness
   d. Incidence of mites, weak desiccant shoots attract mites
   e. More stress by drought, and weak bushes are more vulnerable
2. Low OM (soil carbon)
   a. Consequential low microbial population results in poor assimilation of nutrient inputs
   b. OM in soil surface serves as mulch, encouraging soft vegetation harboring beneficial predators and parasites, reducing the soil moisture evaporation (moisture loss)
   c. Build-up of beneficial microbes Trichoderma, Pseudomonas, Paecilomyces, Metarrhizium, and mycorrhiza while suppressing/preventing other pathogenic ones.
   d. Humic acid generating from OM is a good source of microbial food and a good buffer, encouraging feeder root proliferation conserving soil moisture and applied nutrients
   e. soil becomes porous aerated, helping drainage
3. Continuous use of herbicides in plantations
   a. Kills the weeds: while killing weeds, affects the tea plants as well, silently eating up the starch reserve and weakening the tea plants
   b. The weaker plants, go through the stresses banjhification, incidence of pests like mites, extending the stress periods adversely

c. The herbicides often get absorbed and translocated and accumulate, recirculate, and get distributed within the system, developing toxic mottling symptoms and scorching. Thus absorbed herbicides remain in the system active at least three months

d. Kills all the soil-borne beneficial microbes, thereby hindering the absorption and utilization of applied manures/enzyme activities

e. Vacancies occur due to mortality of weaker bushes reducing the productivity and quality; triggering incidence of pests and diseases escalating the cost of plant protection. Thus the adverse effects of soil low pH, low OM, low microbe count, weaken the bushes, reducing the root starch reserve below threshold, and forcing the shoots banjhi. In young tea, the laterals produced from the main stem scorched and dead, delaying the ground cover. In sum up, every drop of herbicide used in tea plantations is a slow poison; causing death of the plants slowly, damaging the environment irreversibly impacting chronic stresses

f. Impacts of Repeated use of, heavy dose of hard plant protection chemicals to manage pernicious virulent pests in the past decade.

   i. The insecticides are not selective but broad spectrum ones. Hence beneficial parasites and predators also get killed, forcing the users resorting on chemical only.
   ii. Repeated use of pesticides at higher doses, contribute for resurgence of more virulent forms, which are not amenable to the conventional chemicals in usage.
   iii. Both contact and systemic chemicals leave residue on the leaves and plant system that can be hazardous to consumers at higher levels. Repeated, higher doses lead to high levels of residues which are objectionable.
   iv. The applying personnel get exposed to the chemicals, predisposing them and children to diseases and unhealthy environment.
   v. The chemicals impart discoloration, bad odor and invariably deteriorate the quality and value realization.

12.4.2 Concluding remarks and suggestions/road map for the well-being of the tea industry of India

The current problems of tea industries world over have been identified and remedial measures with time frame and packages for ecofriendly GAP/natural farming have been evolved with dependable Technology and proven cost-effective inputs and practices.With initiation of recommended field practices and nature's cooperation, gardens emerge with good satisfactory crop response and distribution with value realization reducing the chemical load, MRL, COP, and so forth, improving the ecology/environment, social harmony, and expectation of the stakeholders. The Prime Minister and governments at the center, and states, are aware of the situation, evolving the strategy and innovative reforms for the revival of the tea industry with commitments from all the stakeholders involved for early implementation and prosperity of the nation, as well as younger entrepreneurs. The ethnic population of the northeast, especially the trained, dedicated, sincere hard working garden work force deserve to be rehabilitated and settled, joining the mainstream with dignity as citizens of a free, democratic India. We are optimistic that the tea industry of India Sri Lanka will revive and flourish and move forward, considering the history of tea development by the pioneers, toil, sacrifices, efforts of establishing the tea, and developing the industry will not go waste. It will revive and flourish with renewed vigor and strength. We the planters never senesce but cherish living with plants and nature. The present technology with cooperation and understanding of all stakeholders will boost the recovery and prosper.

The objectives of the packages and the calendar of operations are to rectify the damages caused by water logging and glyphosate sprays; restoring the health of the soil and productivity of the bushes, through ecofriendly way reducing the chemical load: Impact of the measures suggested and implemented:

1. Soil health improves-pH optimized, OM level, Microbial population, Feeder root proliferation are restored with good response to all manurial inputs in the soil as well as foliar, triggering the metabolic activities of the plant; reflecting on sustainable productivity with quality.
2. Because of the improved soil and bush health, response to all inputs will be good reflecting on the COP, saving the manurial and plant protection costs as well as the economy.
3. Bushes will respond to the efforts and inputs applied, based on scientific principles.

12.4.3 Productivity and quality of North Indian tea-quick regeneration package and time frame envisaged, 2020–22

1. Current position and causes for the stagnation with poor value realization
2. Remedial/Restoration measures proposed/suggested

3. Calendar of operations/Schedule of action
4. Expected outcome/response/impact by the year 2022
5. Achievements anticipated and follow-up proposed to reap the benefits of efforts

12.4.3.1 Calendar of operations (North India)

1. Blanket application of dolomite @ 1 Mt per Ha to optimize the soil pH—January 2020
2. Addition of Enriched Om@100 kg per Ha—January 2020
3. Consolidation of both shade and tea;
4. Reduction and withdrawal of herbicides from plantations
   Reduction to 1.5 rounds per year—2020
   Reduction one round per year—2021
   Complete withdrawal/phase out—2022
5. Initiate ecofriendly package (IPM, bio + conventional): Continue for three years and confirm protocol, recommend GAP/green 2023
6. Amend, revise, and upgrade, Plantations/Labor Act 1951, 1956 to facilitate the mobility of younger generation to join the main

Stream taking advantage of the welfare programs of both state and central schemes.

Prime Minister's outreach (central) programs may be launched as a special care of women workers and young school-going children. Review outcome of the measures, follow-up steps taken.

12.4.3.2 Impacts, intricacies, correlation, coordination of soil/plant biospheres, and biodiversity on dynamic physiological metabolic processes, thereby productivity, quality, value realization, economics, sustainability, and the longevity of tea plantations

Soil, Plant, environment, Biodiversity and Finance of Tea Plantations play a major role on health, productivity, quality, sustainability, cost of production, value realization, Economic and Longevity/life span of plantations. These principal factors are interrelated, correlated, coordinated governing the growth and welfare of the plantations and the depending population, the work force, management and the stake holders. Any deficiency or deviation from the fixed norms will affect the performance of plantations.So care should be taken in planning and achieving/optimizing the set parameters implementing the Management Practices on time to achieve the target/goal successfully. The Expected goal and time frame for achieving 5000 kg/ha made tea with satisfactory value realization minimum price Rs 500/kg, adopting ecofriendly GAP/natural farming are narrated below for the industry management:

1. Soil Profile and Biosphere: Soil profile besides climate/environment decide on the suitability of crops stress tolerance, spectrum of vegetation, growth and development. An ideal soil for tea and related Horticultural intercrops need to be of 3m deep, for anchorage and nutrient, moisture exploitation with the top soil of sandy loam nature, with high OM,moderately acidic (pH 4.5—5.5) without harmful heavy metal contamination. This will encourage facilitating the build-up of beneficial microbial population, feeder root proliferation. Steep slopes, and marginal soils can be avoided and diverted to suitable horticultural crops.
2. Plant/Cultivar: Elite planting materials of both seeds and clones should be selected and planted out at 15,000/ha besides reclamation and revival of present old plantations. The bush architecture, pruning, harvesting, time and type etc should be fine-tuned, upgraded for mechanization and labor saving using the technological advances.
3. Climate/Ambient environment: The ambient environment plays a significant role on the metabolic functions of the plants already established. Location-specific nixed shade of legume trees,cover crops, wind barrier etc should be established/restored for optimizing the parameters for efficient photosynthetic, metabolic processes, enhancing the growth and development of the plants and the workers efficiency. Similarly the moisture content and aeration in the soil also play a key role, on the biospheres regulating the growth and development of plants. The prescribed parameters should be optimized for sustained productivity with quality; fine-tuning the same to the local terrain, climate, cultivars, end product, judiciously by the Management streamlining the inputs, adopting the good Agricultural Practices/Natural Farming.
4. Important Field management practices for Established plantations:
   a. Consolidate the bushes and shade tree population filling up the vacancies
   b. Reclaim the soil with inputs like soil amendments, organic materials—biomass,cattle manures composts, oil cakes, vermin wash with cow dung slurry, bioinputs like biocides, Bioconsortiums, Bioplant growth promoters for enriching the soil with humic acid, microbial population strengthening the root system enhancing the health and productivity of plants with stress tolerance.
   c. Flood control, waterlogging/stagnation, Drainage, soil and water conservation measures

d. Pruning and Plucking/Harvesting
e. Inputs time and dose, application Rationalized, Balanced, yield prediction, Ecofriendly.
f. Processing- Innovations, automation, Fine-tuning based on current technologies
g. Diversification, Value addition and aggressive/effective marketing.
h. Emphasis and Concentrate on vertical growth rather than Horizontal
i. Present Problems and Remedial measures suggested on priority for revival.
j. Ensure ecofriendly GAP/natural farming avoiding usage of all hard chemicals in the interest of environment, consumers, biodiversity and Nature.

5. Certain Important queries and Clarifications:

a. Age/Longevity of tea plants: Very often planters assume that the economic age limit for tea plants is 50years suggest uprooted or rejuvenated with infills after 50 years. It is not true. By habit the tea plants are perennial trees and have no age limit just like any other perennial trees in the world jungles and forests like Amazon tropics, and California, Sequoia Red wood trees. The dicot trees are vested with cambium in the vascular bundles which generate new layers of growth every year strengthening growth and vigor. Unless the trees are exposed to biotic, abiotic stresses which will weaken them. The weak plants are more vulnerable and succumb causing vacancies. The tea plants are forced vegetative into shrubs by periodic pruning and continued harvesting of new emerging shoots for making the commercial tea. In addition the tea plantations adopt intensive cultivation with chemical inputs that aggravates further. The living China tea tree of Yuvong valley in China is a convincing example for longevity of tea plants. If the plants are maintained under optimal environment and balanced growth (soil/plant biospheres), their longevity/life span bound to increase. There would not be any need for rejuvenation subsidies. Even if rejuvenated, the payback period can be hastened/ curtailed to 3 years, modifying the soil reclamation/Rehabilitation procedures through new available technology and inputs. This aspect deserves to be explored and decision taken. Marginal substandard, slopy soil with chronic pests and diseases, drainage, drought etc can be diverted to other suitable horticultural crops. Future expansion will be on vertical growth supported with suitable amendments in plantations act.

b. Weed management: Pioneers recommended clean cultivation, blanket eradication of weeds. Of late this encouraged the usage of herbicides in plantations. After realizing the impacts of irrevocable damage and chronic stresses by continued use of herbicides both contact and systemic, sparing of soft herbal grass weeds under the canopy is encouraged for soil moisture conservation, microbial population build-up and effective nutrient absorption. Use of any herbicides is prohibited in young tea plantings: Replantings and rejuvenated infills encouraging the early ground cover. Integrated measures suggested/advocated in mature tea, avoiding the usage of herbicides in the interest of environment, consumers, work force and local population. Time has come for withdrawal of herbicides and hard chemicals use in plantations/horticultural crops, finally banning import or production of the harmful hard chemicals limiting the MRL.
Chemical companies like Monsanto of USA, ICI of UK, and Bayer of Germany are the major producers and suppliers of Agro chemicals worldwide. Though they helped the green revolution and food security, time has come to divert their expertise toward production and supply of safe chemicals/inputs for the welfare of the population/humanity the world over (the orders of the judiciary of San Francisco, California, USA, 2019–20).

c. Productivity: Intensive cultivation like staggered pruning, increased bush population and nutrients NPK inputs without balancing with organic bulk inputs or soil amendments aggravated the situation in the fields. The tea bushes became weaker and exhausted, going through the chronic stresses exposed to resurgent pests and diseases, warranting use of hard chemicals repeatedly impacting on the productivity and quality. These adversaries/induced stresses in turn caused defoliation, dieback, vacancies of both tea bushes and shade trees increased cost of production, affecting the economic viability of plantations/bankrupt.
Though the situation was alarming, not adequately equipped with technology, package of practices and inputs, continued efforts, favorable rainfall, availability of bioinputs, initiating GAP, optimizing the soil, plant biospheres, new reformed package of practices have come handy and the tea industry is back on the rail toward prosperity with adoption of GAP/natural farming, diversification, and value addition.

d. Nutrient management: It is one of the principal factors deciding the health and productivity besides the cultivars, soil, plant, water, shade, pruning plucking, inputs time and dose and method of application, bush architecture, population, drainage and related, stresses in the field and management practices. Optimum dose is based on the estimated crop, replacement theory, balanced, fixing the ceiling to avoid the toxicity and adverse effects on micronutrients, soil reaction, health and quality of the produce; overdoing anything is poison. Field trials indicated that the optimum 100 kg/ha nitrogen balanced, with equal quantity of organic source at 2:1:2 ratio applied in three to five splits depending on the crop distribution.In field situations of terrain, climate, almost 50% is lost due to volatility and leaching. Ultimately about 30%–40% is available to the plants. Their absorption efficiency is around % depends on the soil, plant biospherrs, and bush canopy, root system, demand and supply governed by source-sink relationship. Flushing behavior and harvesting also play a role. It has been found that about 11%–12% of N is retained in maintenance leaves/shoots at the year-end sustaining the metabolism moderating/of 2000–3000 kg productivity provided all the prescribed soil, plant, shade parameters meeting the stress periods. It is obvious that the bushes can sustain the minimum economic productivity of 2000–3000 kg/ha at 100 kg nitrogen, balanced with,organic sources 50:50 applied in three to five splits at 2:1:2 NPK ratio, provided all the prescribed soil, plant, shade, environment, and parameters are optimized and met. Any deficiency or overdoing is harmful, impacting on the productivity, health, economy, and life span of the bushes.

e. Plant Protection: Usage of chemicals and cost of management—considering the MRL and deterioration of quality of made tea reducing the value realization, this is one of the areas of serious concern allover tea plantations at present. However, consequent to the realization and implementation of GAP/Natural farming, availability of dependable proven quality bioinputs and technology/packages, strengthening the soil/plant biospheres, consolidation of shade,-boosting the plant health, growth, and development have improved the field situations/conditions, with satisfaction gaining confidence. The plants are exposed to least stresses free from any serious maladies, pests and diseases. Thus the cost of production and chemical load have been significantly reduced, hiking the value realization. It will be a happy win-win situation for all the stake holders. The current sale price of made tea in international market reflects good future for the tea industry on adopting GAP, diversification, and value addition.

6. Future of tea industry: priority areas of R&D areas suggested:
The future of tea industry is good. Revival and prosperity are ensured, considering the back ground history, efforts, steps taken by the industry, Managements and Government in the interest of all stake holders-millions of depending work force, local population, consumers, economy of the Nations.

There are positive valuable scientific information that there is enormous Scope for improvement in plant science using genetic mapping, breeding for crop improvement, Microbiology and Biotechnology for Organic inputs replacing hard chemicals, both biocides and biofertilizers inventing nano micronutrients and bioplant growth promoters accelerating the Photosynthetic capacity of leaves thereby, productivity, quality,the longevity of bushes and cost-effective.The R&D infrastrucutres facilities, and scientific personnels should be created and introduced, Reviving the Pioneer world's reputed Tea Research institute at Tocklai, Jorhat Assam, India as an international tea research and horticultural food processing, catering the need of world tea/horticulture making the best use of the technology and wealth available, for the welfare and health of humanity. There is enough scope for cooperation and coordination of positive exploitation of available technology, developing ecofriendly GAP/natural farming packages of practices for implementation in the interest of food, nutrition security, and safety Completion of the northeast development infrastructure, horticulture and dairying projects, northeast will be a potential area for tourism.

a. Control of recurring floods in the Brahmaputra river; stabilizing the flow, water level from ground with measures like dredging, restoring free down-flow, preventing inundation and water logging in Assam, Arunachal plantations, restoring the high quality productivity without chronic recurring stresses. This will reflect on the sustained increased level of productivity with quality, value realization and cost effectiveness.

b. Implementation of Prime Minister's North East Development Projects.

   i. Horticultural project and poor family dairying and fishing, including procurement, storage, processing, and marketing in the mighty river banks and valley resettling the surplus workers of tea plantations due to mechanization, duly imparting skill training and resettlement in the Assam valley. This will be complimentary to plantations, supplying the sought after, organic biomass, cow dung, urine, compost and cattle manures facilitating the GAP/Natural farming envisaged, creating employment opportunities uplifting the tribal population joining the main stream, contributing to the prosperity of the Region. This region can be developed in to a tourism area for the visitors, promoting the Tea sales through Show rooms.Both small tea growers and corporate can be encouraged to participate and gain.

ii. Welfare Measures for the plantation workers of five generations, in Assam, Arunachal and Doars, Terai, Darjeeling in West Bengal settled and merged with local population, facilitating their joining the mainstream. Their minimum wage should be enhanced/fixed linking with productivity, housing facilities, education and health facilities for the women and children. Housing and land for cultivation could be provided on lease by the governments, to provide the food security, self-employment, and sustainability of the displaced workers as the citizens.

iii. The northeast is the paradise of India with enormous natural wealth and resources. This can very well be utilized for the prosperity and well-being of the people and the nation at large.

iv. The Plantation and Labor Acts of 1951 and 1956 need to be amended, facilitating growing of proven local horticultural crops of the region for diversification and value addition by small growers as well as corporate interests.

7. Foliar application of micronutrients and plant growth promoters regulators.

The negative mindset of many oldtimers incuding the advory is broken with advanced technology and bioinputs available. It has been proven in varied field situations and convinced the planters, that these nano micronutrients, bioinputs are potent tools for crop promotion without any side effects but boosting the growth and productivity of the tea bushes. Need-based, discretionary use of these valuable inputs alone or in combinations/consortiums boost the crop up to 30% seasonally, nevertheless up to 10%–15% increase with assured cost benefit ratio 1:5. These bioproducts and nutrients can also be used with irrigation sprinkler for stress removal in winter Figure 12.1.

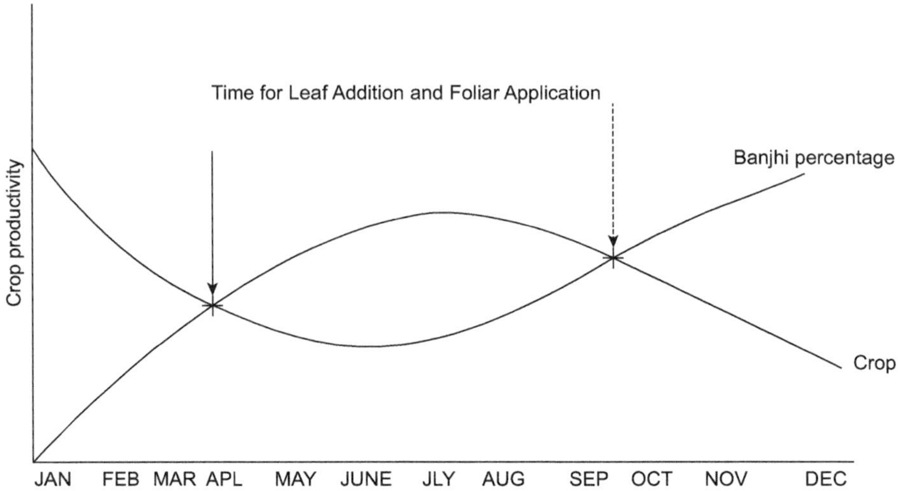

Figure 12.1 Time for Leaf Addition and Foliar Application for crop regulation.

12.5 Concluding remarks and suggestions/road map for the well-being of the tea industry of India

The current problems of tea industries world over have been identified and remedial measures with time frame and packages for ecofriendly good agricultural practices (GAP)/natural farming have been evolved with dependable technology and proven cost-effective inputs and practices. With initiation of recommended field practices and nature's cooperation, gardens emerge with good satisfactory crop response and distribution with value realization, reducing the chemical load, MRL, COP, and so forth, improving the ecology/environment, social harmony, and expectation of the stakeholders. The Prime Minister and governments at the center, and states, are aware of the situation, evolving the strategy and innovative reforms for the revival of the tea industry with commitments from all the stakeholders involved for early implementation and prosperity of the nation and younger entrepreneurs. The ethnic population of northeast, especially the trained dedicated, sincere hard working garden work force deserve to be rehabilitated and settled, facilitating to join the mainstream with dignity as a citizen of a free, democratic India. We are optimistic that the tea industry of India and Sri Lanka will revive and flourish and move forward, considering the history of tea development by the pioneers, toil, sacrifices, efforts of establishing the tea, developing the industry will not go to waste. It will revive and flourish with renewed vigor and strength. We the planters never senesce but cherish living with plants and nature. The present technology with cooperation and understanding of all stakeholders will boost the recovery and prosper.

CHAPTER THIRTEEN

Case studies, field observations, and troubleshooting

Contents

Scientific Perspectives of Tea Plant Horticulture and Productivity
ISBN: 978-0-12-823444-0
https://doi.org/10.1016/B978-0-12-823444-0.00013-3

13.1 Introduction

The field performance and productivity of plants are influenced by various factors. On the spot inspection, observation and timely action to rectify and ameliorate,minimize the damage/impact,restore the health and productivity. Most of the recommendations /suggestions are based on the field demos and observations, applying the expertise, field situation and cost effectiveness.

13.1.1 Usage of bioproducts in tea plantations for sustainable economic productivity

Tea production is sustaining improving trends, despite all the constraints and vagaries of monsoon and competition in the world market. The image of Indian tea being a cleaner produce received a setback due to outbreak of new strains of virulent loopers, mites, and tea mosquito bug in tea plantations, warranting repeated use of higher dose of hard chemicals including synthetic pyrethroids. Consequently, resurgence of resistant pests emerged which are tolerant to the conventional methods of control. Labor shortages in plantations has also added to the problem leading to mechanization and repeated overuse of herbicides. As a result, cost of production escalated and the quality of the product deteriorated affecting the image and value realized in the international arena. Now, time is ripe for introspection and hard decisions by all the stakeholders of the Indian tea industry to restore the image and glory of our tea as the champion of world tea.

13.1.2 Suggested ways and means

The five cardinal points for the revival program are the following:

1. **Attend to the basics** and stick to the timeframe scrupulously following the Tocklai/UPASI recommendations and improvements
2. **Optimize** the soil and field conditions for congenial productivity with quality
3. **Initiate good agricultural practices**, minimizing the chemical usage, adopting conventional ecofriendly field practices preserving nature
4. **Effectively harvest the crop and manufacture, market judiciously**
5. **Explore Rationalization of inputs, Energy saving, Mechanization, value addition for Economy and sustainability**

13.1.3 Practical suggestions for implementation

- **Consolidate and fill up the vacancies** with improved planting materials
- **Restore effective drainage during monsoon and adopt drought amelioration measures during winter**

- Conserve organic matter and enrich the soil with compost for aeration, feeder root proliferation and microbial-build up effecting economy of inputs (nutrients)
- Minimize the usage of herbicides to sustain the microbial population, soil and bush health, saving valuable NPK inputs retaining the quality of the produce
- **Encourage liberal use of** biofertilizers such as **Azospirillum, Phosphobacter**, Vesicular Arbuscular Mycorrhiza (VAM), and K solubilizer along with compost as first split for the season to effect economy of NPK fertilizers—up to 40% saving can be achieved in a pruning cycle
- **Encourage use of** bioproducts (Trichoderma, pseudomonas, and phosphobacter/VAM)in the planting pits, on pruned frame(Trichoderma + B. subtilis), foliar spray of UP fields in the winter with Bioconsortium (Beauveria+ Metarhizium + verticillium + Paecilomyces + pseudomonas) against the sucking insects, loopers, and termites, encouraging the predators, thereby reducing the number of rounds of chemical spraying during the crop season restoring economy and quality
- Adopt drought amelioration measures in YT and UP mature tea with leaf addition and foliar of antitranspirants (NK, Mg) and osmoticum combinations at fortnightly intervals commencing from January to March to promote health and productivity of the bushes
- Use biodigester for production of compost and preservation of pruning litters wherever feasible, establishing wind barriers, cover crops, and recycling the farm wastes
- Winter operations like drain cleaning, hoeing, leveling for YT and pruned fields, NK foliar for UP fields and drought prone fields, and Level Off Skiff (LOS) for helopeltis affected fields are mandatory
- Judicious use of proven growth promoters combined with micronutrients is recommended wherever stagnation of crop is evident for smoother crop distribution.

13.1.4 Specific recommendations for immediate decision

- Commence NK+GHOM foliar fortnightly in all the UP fields, dose being-2 kg urea, 2 kg MOP+ 100 ml per drum per ha. In case of helopeltis attack, level skiff the area with mechanical harvesters and give the foliar during first week of February.
- Apply Bioconsortium foliar in one section each, of TPS, Gungaram, DemDima and Killcot and monitor the outbreak of/control of pernicious pests like helopeltis, loopers, and termites during the peak season. Based on the impact and economics decision can be taken.

- Similarly apply Azospirillum 5 kg, Phosphobacter 5 kg, VAM 5 kg+K solubilizer 5 kg mixed in 50 kg of compost and broadcast in selected UP fields during February/March on receipt of first rain, and see the response. The NK dose can be pegged to 150 kg per ha instead the 180 kg irrespective of the yield level.
- Use of bioproducts in tea plantations of Doars and Darjeeling.

Name of product	Purpose of use	Dose/mode of application	Remarks
Azospirillum	N fix fertilizer	Planting pits@ 20kg/ha	Mix with OM & apply
Phosphobacter	P releasing bacteria	do-	do-
Vam	P mobilizing fungus	do-	
K-Mobilizer	Bacterial culture	do-	
Trichoderma viride	Fungal antagon	10 kg/ha	Mix with OM, broadcast in moist soil
Beauveria bassiana	Fungal insecticide	5–10 kg/ha	Against Helopeltis, Shothole borer, loopers
Paecilomyces	Fungus culture	10 kg/ha	Nematicide, works against thrips with BV
Verticillium	Fungus culture	-Do-	Acaricide apply with jaggery
Metarrhizium	Fungus	!5–20 kg	Used against termites & mites apply with molasses
Biodigester	Mix of Fungus—Aspergillus	20 kg/ha	Apply on prunings, compost making
EM4/6	Effective microbes	Multipurpose 10 kg/ha	
PGPR	Microbial stimulants (probiotics)	5 kg/ha with starters	Used as synergists with nutrients/humates for stress tolerance and crop growth as booster

- Fine tuning of dose and timing will be required based on the experience and field situations. Try in divisions of problematic gardens for demo and decision.

L. Manivel
Consulting Scientist, Plant Physiologist
Coimbatore, Tamil Nadu
India
Dated, December 29, 2013

13.2 Current problems and probable solutions of tea plantations of North East India

13.2.1 Introduction

The status of the tea plantations of Assam, and Bengal in particular during end of the century as compared to the midcentury, would reveal the drastic changes the industry and plantations have undergone, consequently contributing to the present problems, in the fields, factory, and marketing. The soil and climate have changed adversely. Unless the soil is enriched and amended, the microclimate intervened through introduction of shade, cover crops, and agroforestry; the future is bleak. Simultaneously, soil, water conservation, and rainwater harvesting should be ensured to overcome the water shortage. Strict adherence and moderation to integrated nutrient management, integrated pest and disease management, including herbicide usage, are required to ensure the cleaner produce in the interest of the producer and consumer countries.

Once the global good agricultural practices (GAP) are enforced and adopted, the spectrum of pests and diseases and weed flora will change for good and the impact and response could be seen with salutary effect. Sociological changes are bound to emerge in plantations.

Emerging new plantings and small scale plantations in nontraditional areas need to be encouraged and guided to produce quality tea under cooperative basis to contribute their might to the economic uplift and progress of the nation. Timely supply of quality planting materials, certified inputs through collective efforts should be ensured and mechanism evolved. As the leaves of these small growers enter the corporate gardens for manufacture, chemical usage, standards of shoots including the chemical residues deserve to be monitored and regulated in the interest of the industry and country.

Mechanization of field and factory operations and thereby the fine tuning of the operations infusing technological innovations and breakthroughs is a necessity for survival. Modernization of factory is an urgent necessity and priority to conserve the energy, economizing the cost of production. Introduction of modern gadgets like electronic nose and electronic vision, will be the added advantage for tea industry to sustain the quality.

Diversification, value addition and aggressive marketing to fetch stable remunerative price for the produce, is a priority area with lot of potential for innovations, harnessing the natural resources of the country. Green tea, organic tea, tea extracts, and tea beverages should be looked into to add value to the current scenarios.

A comprehensive, package of guidelines making available the best possible tea to the consumers, harmonizing the efforts of agencies involved should achieved early, in the interest of the stakeholders of the industry.

13.2.2 Current problems of serious magnitude and probable solutions

The problems are listed below in a descending order of their seriousness requiring attention on priority.

Soil and environment: Initial plantations were planted out in virgin soil with plenty of humus with optimum soil reaction, clearing out the dense forests. Consequently, the soil was rich with high organic carbon, enriched with microbes, with adequate nutrients to sustain the plant growth for sustained productivity.

The situation changed radically now, due to intensive monoculture, changing course of rivers, inundation and inadequate outfall, encroachments and widespread deforestation, leading to change of monsoon patterns, weather and climate change, erosion and aggravated floods.

Change of field management practices, spectrum and population of planting materials, age of the shade trees and tea bushes, compounded with maladies like storms, hails, aggravated by pests and diseases resulted to the present adverse conditions of plantations affecting the productivity, quality, and value realization threatening the viability of the operations/plantations.

13.2.3 The key factors which deserve immediate action to stem the rot are the following

13.2.3.1 Optimize the soil reaction through application of appropriate soil amendments

Liberal use of fertilizers to harvest the maximum crop every year, affected the soil reaction adversely reducing below the threshold level of 4.5. Similarly, soil reaction escalates beyond 6.0 in riverbeds due to inundation and flooding. Consequently, the absorption and assimilation efficiency of applied nutrients decreased. Certain essential micronutrients become unavailable under adverse soil reaction. Low soil pH is one of the serious problems in plantations that is responsible for stagnation and deterioration in quality. It will be prudent to apply 1 MT of dolomite per hectare at the time of pruning and incorporated, to rectify the soil reaction restoring the productivity. Similarly, iron pyrites or gypsum can be used to lower the soil reaction in high pH areas. Calcium, sulfur, and magnesium besides correcting the soil reaction, serve as valuable secondary nutrients contributing to productivity abundantly.

13.2.3.2 Consolidation of shade and tea, filling up the vacancies caused by different factors duly correcting the predisposing factors

The vacancy percentages of both tea as well as shade are high (around 20%) due to adverse soil and high water table, weakening the plants, in turn inflicted by the root diseases. Until the vacancies are rectified/consolidated, the crop loss is imminent. Both shade and tea have to be infilled at the time of medium prune/HRD or rejuvenation once in 15–20 years as a policy, to restore and sustain the productivity.

13.2.3.3 Buildup of carbon content of the soil encouraging the population of the beneficial soil microbes

Introduce, modify and streamline the shade pattern, cover crops, binding grass implementing soil conservation measures, etc., even in hills like Darjeeling for sustained productivity with quality. Wherever the carbon level is too low, warranting restoration oil cakes, composts, and organic manures, can be mixed with soil amendments as well as fertilizers and applied/incorporated. Under the changed current situations of soil and climate, the intensity of shade, spectrum of shade, regulation, and need of shade in hills require a second look to optimize the photosynthesis, nutrient use efficiency for sustainable productivity with quality.

13.2.3.4 Establishment of wind barriers, fuel forestry in the peripheries and marginal lands to augment the fuel requirement and build up the microclimate

Reduce the dependence on chemicals for plant protection including weed management, restoring the soil health. Self-sustenance of mulching material, compost, and organic manures should be encouraged for cost-effective young tea management.

Streamline the water management in the region and thereby ensure the availability of adequate water throughout the year, harnessing the water resources and regulating the distribution and utilization. Rainwater harvesting, control of soil erosion, power generation, and flood control are some of the priority areas involving interstate, center state coordination, and enforcement.

13.2.3.5 Strict adoption of soil conservation within the gardens and afforestation, making use of the surplus, marginal land available in the plantations

Introduce temporary shade, cover crops, binding border grasses, live fence, cover crops, liberally, lop periodically and enrich the soil

Diversify with value added intercrops, in the plantations without hurting the main crop

Introduce selectively the aquaculture clubbing with the water conservation measures.

Introduce dairying, poultry, and goat rearing, horticultural crops, utilizing surplus marginal land for viability, self-reliance for inputs, nourishment and economic viability through mixed integrated approach in plantation management.

13.2.4 Field management practices—probable solutions are enlisted below

13.2.4.1 Young tea

Under the changed soil and climatic conditions, depth of planting pits, inputs in planting pits and after care are essential for better establishment with quick ground cover and early crop. The guidelines for young tea management should be strictly followed and enforced especially under the present tendency to skip the rehabilitation, planting done in nontraditional areas using marginal soils. As the microbial population declined, availability of nutrients to the roots declined, OM level reduced, soil is marginal, climate is hostile, the pits/trenches should be deeper, enriched with compost, rock phosphate, VAM, *Trichoderma* etc. should be liberally applied in the pits. Hardy, grown up healthy plants should be preferred as planting material. Besides Tocklai and garden released clones, proven Biclonal stocks should be encouraged as planting material for the sake of plasticity. To conserve the soil moisture, mulching should be ensured. Cover crops and shade establishment-both temporary and permanent, should be ensured. Usage of herbicides during first year should be discouraged and moderation ensured especially in use of translocated systemic herbicides like glyphosate until formative prune (FFP1).

13.2.4.2 Mature tea

The desired soil status such as, soil pH, organic matter content, porosity, and breaking the soil pan through forking, leveling, shade stand, adequacy of drainage, correcting outfall problem, correction of frame height through HRP, repairing the frame, infilling both shade and tea filling up the vacancies, should be attended periodically (once in 15–20 years) to revive the root system; restoring the balance between root and frame. The HRP fields require close attention in terms of soil amendments, OM addition, hoeing and leveling, tipping, canopy management for a minimum of two pruning cycles to restore them normally. The drainage and potash status of the soil and effective removal of excess water from the root zone are important to overcome the algal parasite red rust (*Cephalurus parasiticus*) and root rot diseases. Incidence of red rust results in defoliation, exhausting the starch reserve, weakening the bush, affecting the productivity and crop loss. Persistency of root diseases weaken the bushes, leading to dieback and vacancies. During pruning time, the cut ends should be protected by application of cocktail of *Trichoderma* + *Bacillus subtilis* slurry along with molasses for protecting the wound/cut-ends.

13.2.4.3 Manuring and harvest of crop

Once the basics are attended and the field conditions, both soil and shade, bush population are optimized, there is scope for reduction, rationalizing the manurial inputs without hurting the productivity or quality, while preserving the bush and soil health for sustainable productivity. The splits can be increased to three to five, coinciding with the bush growth phase. Organic manures such as compost, cattle manure, and oil cakes can be included along with the fertilizers during first and last splits for economy, efficacy, and better utilization of the inputs.

Considering the labor availability, effective timely harvest of the crop is going to be the challenge. Resorting to mechanical harvest is the only answer. In order to maintain the health and productivity of the bushes, periodical (2–3 months interval) foliar spray of nutrients supplements should be ensured, to prevent exhaustion and incidence of gray blight (Pestalotiopsis).

13.2.4.4 Plant protection

The spectrum of pests have changed during the past decade due to the changed climatic conditions. Consequent to the occurrence of pernicious virulent types of pests like loopers, helopeltis, and mites, leading to repeated use of hard chemicals like synthetic pyrethroids, management of pests becomes difficult escalating the cost of management. Resurgence of sucking insects evident, dominating the scene, persisting throughout the season. Moderation in use of chemicals, intervention of herbal extracts, biocides encouraging the natural enemies (predators), and adopting IPM modules is an urgent necessity. The available information in these aspects should be scaled up and guidelines issued.

13.2.4.5 Diversification, value addition in manufacture

Production and supply of consistent quality product at affordable price to the consumers is important and paramount for the future growth of the industry. Energy saving in manufacture, innovation and introduction of modern technology in machineries, processing technology besides diversification in to green, herbal, organic, packeting are urgent for the very viability of the industry.

13.2.5 Thrust areas research for the future of the industry

- Development of elite planting material combining the traits of stress tolerance, productivity with quality
- Evolution of package of practices and guidelines for economic, eco-friendly management of pests and diseases
- Development of nutrition requirement and package of practices (INM) for sustainable productivity with quality under varied soil and climatic conditions
- Develop alternative ideal shade species for the field situations and optimize the shade requirement under different terrains, regions for sustainable productivity with consistent quality
- Development of tea processing machineries and field appliances for cost-effective management
- Producing cleaner produce with consistent quality
- Development of systems for soil and water management for the entire northeast including Darjeeling hills

References

1. Encyclopedia of Tocklai, TRA, Jorhat, Assam, 785,008.
2. Peter, F. (Ed.), 1975. Hand Book on Tea Culture.
3. Anonymous, 2011. Field management in tea. In: Hazarika, M. (Ed.), Tea Research Association, Tocklai, Experimental Station, TRA, 2012 pp.
4. Manivel, L., 1999. Physiology of tea productivity. In: Jain, N.K. (Ed.), Global Advances in Tea Science. Aravali Books International (P) Ltd., India, 1999. pp. 463–488.
5. Manivel, L., 2011. Past, present and future of the tea industry-a perspective of R&D need for the century. Concourse, Souvenir of World Tea Science Congress. In: Hazarika, M., Baruah, P., Barthakur, B.K., Chaturvedi, S.K. (Eds.), Celebrating 100 years of Excellence in Tea Research, 2011. Tocklai Experimental Station, Tea Research Association, Jorhat. pp. 73.

L. Manivel
Consulting Scientist, Plant Physiologist
Coimbatore, Tamil Nadu
India

13.3 Drought amelioration during winter: Mackaibari TG, Kurseong, Darjeeling district

Mackaibari being the organic garden, only ecofriendly bioorganic measures are suggested for the management of the drought during the winter. These measures are expected to preserve the health of the bushes, assisting the bushes to recover early from the stress (drought), and producing the quality early crop, free from pests and diseases, paving the way for early quality crop boosting the economic viability. The following are the measures narrated for guidance:

1. Add a new leaf in the canopy filling the pockets and liberal plucking. A recently mature leaf in the canopy is a good moderator and insurance against drought than old mature leaves or tender leaves. Leaf addition and strengthening the canopy should be completed before the onset of winter, the latest by end November.
2. Fortnightly foliar of NK(Potassium Nitrate @1%)Alternated with Potassium Sulfate, plus cows urine, cow dung slurry fermented for 3 days along with proven herbs like, polygonum, clerodendron, cassia alata, vitex, Adathoda (Thithaphool) seethaphal leaves etc.), Potassium

sulfate plus synkrovalar at 100ml per drum(stress amelioration dose) to serve as osmoticum controlling the transpiration water loss.

3. Neem kernel extract (NKE) slurry to control the sucking insects common in the winter for preserving the bush health especially in UP fields. The procedure:
   Take 5 kg of neem seeds crushed, plus 1 liter of cow urine, 1 kg of cow dung, 3 liters of water, make slurry and ferment for 3 days, decant the supernatant, dilute five times, about 100 liters water and spray covering the tophamper per ha. All the sucking insects including the Helopeltis will be under control with this NKE foliar.
4. Soil Mulch to break the capillary transpirational loss of moisture, reducing the weed competition.
   Mix the OM of Synkrovalar 500kg per ha + vam, Trichoderma @ 5kg per ha, broadcast scrape/hoe and incorporate. This measure will help in adding up Om enriching the soil, arrest the capillary water loss, weed competition while aerating the roots, protecting the roots from the pathogens(Root diseases).
5. Foliar of herbal extract of Clerodendron, polygonum, *Cassia alata*, seethaphal, *Vitex negundo*, adathoda, and so forth, extract in cows urine, dilute, and drench the canopy to impart drought tolerance as well as plant protection preserving the health.
6. Feb: give a foliar of 2 ppm Triacontanol (Vipul or phytonol) plus 1% synkrovalar 100ml per drum to impart drought tolerance, restricting the photorespiratory biomass loss.
7. Spray Trichoderma Plus B. subtilis cocktail spray @3+2 Kg per ha in 200lit water on the pruned bushes to protect the cut ends, hastening the wound healing, and early break.
8. Check the soil pH through soil analysis and amend/optimize the soil pH through soil amendments like dolomite, sulfur or gypsum, magnasite bioproduct containing Ca,5.5% Mg 8.5% + S 31% (source from Salem Magnasite factory @ 25/kg. Sample can be arranged for demo.
9. Feb: Level up the canopy and give a foliar of synkrovalar @ 200 mL per drum for early bud break.
   Apply Synkrovalar/Yazhisri Bio OM, Chennai, broadcast and incorporate @ 500 kg/ha. Raise leguminous intercrop like, Arhar, Sun hemp, Daincha, sesbania, etc in alternate rows as cover crop, slash and incorporate before flowering adding OM to the soil serving as green mulch.

10. March: Foliar of Synkrovalar+ KNO3, followed by a spray of Bio-consortium against sucking insects including Helopeltis, thrips and green flies.

L. Manivel
Consulting Scientist, Plant Physiologist
Coimbatore, Tamil Nadu
India

13.4 Visit report of Boisahabi TE, Jorhat, Assam, February 20 and 21, 2019

Joint Visit by, Dr.L. Manivel, Plant Physiologist, and Dr. P. Tamizh-chelvan, Agri Microbiologist, Vishakan Biotech, Coimbatore 641046

Summary: The Boisahabi TE with Boisahabi and Hemla divisions about 500 ha, located on the high way between Mariani and/Tinsukia, adjoining the Nagaland border. The terrain is flat with clayey soil and intermittent streams and rivulets, resulting in high water table during monsoons. Most of the plantations are raised without rehabilitation and trench planted. The garden is ecofriendly, maintaining their own dairy, of about 20 Jersey/Yorkshire cows for microbial fermentation of slurries for INM and IPM.A gas plant also in operation for cooking gas.TV23 seems best suited for the region touching the productive level of 3500/kg/ha, with the population of 13,500–15,000/ha.

Albizzia mollucana is the shade in the entire garden, maintaining the shade with periodic pollarding. Thus they could meet the firewood requirement for worker, while building up the carbon/humic acid level in the soil. The ecofriendly cultivation of tea in this garden reflects on the biosphere, environment, bush health, soil health, productivity with quality (documented with crumply soil with feeder roots, earth worm activity, plenty of birds in the fields morning and evenings, absence of pernicious pests or diseases, bringing the chemical loads to the minimum, and fetching a reasonable price for the produce).

The principal problems faced by the garden are:

1. Persistency of red rust for prolonged period April–July
2. High water table restricting the root system of both tea bushes and shade trees to top soil of 45 cm
3. Prolonged inundation and stagnation of water in the root zone, restricting the aeration, affecting the root proliferation and microbial population impacting on the absorption efficiency of applied inputs for the plants in

the soil, in turn affecting the productivity and quality and cost effectiveness. A ratio of 60:40 soil moisture and air ideal for plant metabolism. Integrated manuring with OM, bios applied in splits will improve efficacy.

4. Despite all the soil and bush health problems parameters like weed growth, mulching, carbon content of the soil, plant metabolism, seem normal a win-win situation for the management and work force in the garden.
5. Agroforestry, fuel forest, wind barrier in the borders/boundaries of the garden and green canopy of shrubs, temporary shade trees in both divisions Boisahabi and Hemla, will go a long way, carrying forward the garden management to the high esteem further.

In addition to the on the spot discussion, clarifications, observations, a few salient points are repeated and briefly narrated for confirmation:

13.4.1 Field-wise observations, comments, and suggestions: 1 Boisahabi division

Sec. No.2,3,4,5,17,12,13,16, and 22 appear to be the areas of inundation and prolonged water logging aggravated with clayey sedimentation of silts from the rivulets, streams, rivers running though. The bushes in these areas are weak, depletion of starch reserve from roots, stressed through lack of aeration leading to scorching, defoliation and dieback, the productivity and quality at the same time escalating the cost of production.

1. One of the serious problems in these fields is the red rust (*Cephaleurus parasiticus*) an algal parasite, occurs concomitant with K deficiency. This had been treated with bioproducts from Vishakan Biotech Coimbatore last year, controlling successfully, from April—July, five monthly sprays. The healed lesions were clearly seen in the leaves. It is time to initiate the spray in the coming months.
2. These chronic waterlogged areas can be explored to drain the excess water during monsoon through some deeper main drains linking the rivulets connecting to a peripheral border drains. The existing pond with lotus can be deepened and developed into a conservation with aqua culture of trouts. Possibility may be explored.
3. Due to the sedimentation, level fields with adjoining paddy lands, prolonged water logging, the pH of these soils invariably at higher level up to around 6.0 coupled with Ca, Mg, and S deficiencies. These type of soils can very well be amended and rectified by one time use of soil

amendments like soil Wonder, a magna site ore product, available through Siva Chemicals, Attur Salem, Tamil Nadu.

4. Introduction of green crops, cover crops, temporary shade trees, lopping and incorporation periodically will improve the soil aeration, OM level root promotion, and so forth, improving the productivity with quality.
5. Diversification with intercrops like fruits, spices, and herbs will improve the property for value addition of quality green tea/orthodox tea produced, through certification and online marketing.

We both enjoyed the excellent organic hospitality, which will be fondly remembered.

L. Manivel and P. Tamizh Chelvan
Coimbatore, Tamil Nadu
India

13.5 Technical discussion with field staff and executives-Halmari, Mokalbari dated Dec. 12th and 15th respectively

13.5.1 Soil health and productivity

1. Soil is the medium for anchorage of plants with roots supporting the livelihood of plants
2. Its structure, contents, status play a major role on health, productivity and quality of produce
3. The principal parameters of concern are
   a. The pH, acidity/alkalinity and Electrical Conductivity
   b. Organic matter content of the soil
   c. Microbial population, flora, fauna that are sustaining the health, Productivity assuring the welfare of the plants (Symbiotic Coexistence)
   d. The moisture, air, and content of the soil plays a role; 60:40 ratio is preferred for satisfactory plant growth
   e. Effective Drainage and out fall, porous sandy loam soil ensure the desired Ratio 60:40. Both Excess and Deficit of soil moisture in the soil lead to many biotic, abiotic stresses of the plants

13.5.2 Impacts of chemical inputs: GAP for tea in Assam

Traditional cultivation practices that were followed using the indigenous resources, cutting a balance between demand and supply, protecting the biosphere and environment. Increased population, advanced technology consequent demand globally, forced the pioneers, reform the package of practices, introducing many innovative chemical inputs which in the long-term became counterproductive. As a result of excessive, indiscriminate use of chemicals, affected the soil health, biosphere, water resources, weakening the plants, predisposing to many stresses. Stagnation of productivity, escalation of COP, affecting the productivity and quality thereby the value realization, making the operation economically nonviable. The solution lies in reverting back to rationalization of all inputs integrating the package of practices, reducing the chemical adopting ecofriendly GAP.

13.5.3 Good agricultural practices for cultivation of tea in Assam

1. Optimize the soil pH, OM level, shade status, plant population, the inputs for sustained productivity with quality.
2. Integrate the fertilizer-manurial inputs @ 50:50 or 70:30 depending on the terrain, environment, cultivar, and productivity. Split application improves the absorption efficiency.
3. Provide adequate shade and wind barriers to maximize the productivity with quality through moderating the photosynthetic efficiency of maintenance leaves. Shade regulation is essential to sustain the Pn efficiency of leaves. Similarly, wind barriers also help against storms and pollination.
4. The input rates are determined based on the population, productivity; it is around 130–150/Kg/ha. Balance 50:50/70:30 between fertilizer and organic manures, if necessary fortified with biofertilizer, biocides and soil amendment (bioconsortiums) to save time and application cost.
5. Grow shade trees both temporary and permanent, cover crops influencing microclimate in the region. Soft weeds in the ground under the bushes are spared and preserved as soil mulch, beneficial to the bushes.
6. Avoid usage of herbicides, fumigants in the soil for weed control, herbicides both contact (paraqat) and translocated (glyphosate) kills the weeds besides burning the starch reserve of roots, weakening the plants besides destroying the beneficial microbes in the soil.
7. Integrated pest management is commenced to save the useful predators/parasites, Environment, producing cleaner quality tea, thereby value realization. Avoid chemicals reducing the chemical load encouraging herbal indigenous softer molecules.
8. Harvest the tea shoots 2+B, weekly (7–8 days), selective periodical, sparing the bud and 1+B for sustained productivity with quality. Yearly at least 30–34 rounds adequate. Both pruning and harvesting play a role balancing demand and supply influencing productivity and quality.

9. Add a mother leaf and in October after Puja holidays to strengthen the canopy helping the tail-end crop imparting stress tolerance.
10. Rest the bushes at least a fortnight before pruning and test the starch level before Pruning. In Assam it should be around 20% for CA/LP.
11. Effective drainage, water conservation measures preserve the health of soil and of the bushes assuring sustained productivity with quality.

Science never fails so the tea bushes. They respond positively through all efforts and inputs.

L. Manivel
Consulting Scientist, Plant Physiologist
Coimbatore, Tamil Nadu
India

List of plants (inclusive) for diversification*

1. Suggested/Recommended species and kind of plants for diversification/value addition
   - **a.** As pump removing excessive inundated water around ponds: *Spathodia, Eucalyptus citriodora, E. globulus, E. robusta, Delanix regia, Enterolobium saman Peltophorrum ferugineum, Glyricidia,* and *Albizia lebbeck*
   - **b.** Wind barrier: Bamboo, silver oak (*Grevillea robusta*), Millingtonia, *Jacaranda speciosa*, Kadam, Saal, Albizzias, Teak, Acacia planiferons, *Dalbergia shisoo*
   - **c.** Fuel trees:Pongamia glabra, Melia azadirach, Shaal, Cassia siamea, Indigofera teismania,
   - **d.** Fruit trees: Avocado, Oranges, Litchi, jamoon, Jack, Areca,Pineapple, Passionfruits,Kiwi Mulberry, jaalakki, Sour Sop *(Annona muricata)*
   - **e.** Herbs.Gymnima, Stevia, mint (Mentha)—for value addition
   - **f.** Grasses. Citronella, Vettiver—for oil extraction
   - **g.** Spices for value addition (inclusive) Cinnamon, Hibiscus, Thulsi (oscimum), Jasmines, jalakki (capscum) lemon
   - **h.** Green crops for shade, mulching, and incorporation in YT and HRP: Sunnhemp (*Crotaleria juncia*), Daincha (*Sesbania aculiata*), Sesbania speciosa/purpurea, Tephrosia, Indigofera, Glyricidia, Gora neem (Melia azadirach), Subabul (Mimosops.), and *Leuaena glauca*
   - **i.** Seed bari for seed production. Helma div.

* The eleven appendices in this chapter carry the salient points of technical information, most of them unpublished, which may be of interest to readers. For reference, contact lmanivel@gmail.com.

j. Bioconsortiums for plant protection including plant growth and development.
k. ROT Arrest, Bio Booster, and Bio Defense—locally formulated

13.6 Field proving of basic package of practices for sustainable productivity in tea plantations- HML-A

Prelude: Tea plantations are more than hundred years old. Soil, climate, cultivar, bush health, spectrum of pests and diseases have changed. Heavy dependence on chemicals is resorted for sustainable productivity, escalating the cost of production. Optimization of the field situations such as soil, shade, and rationalization of inputs through ecofriendly (GAP) measures need to be proven and fine-tuned in tea plantations for area specific recommendations.

13.6.1 Introduction

South Indian tea plantations are located in western ghats hills benefited by the south west monsoon. The soil is of lateritic and is low in organic matter status due to continuous use of heavy dose of chemical fertilizers without adequate amount of organic carbon. The adopted shade *Grevillea robusta* is a nonleguminous plant. Due to slope, soil wash, exposure, inadequate carbon only about 30–40% of the applied fertilizers are absorbed and used. Thus the high dose and split application escalates the cost of production, adversely affecting the soil pH into hyperacidic. Under these adverse soil conditions of low pH, low organic carbon, low microbial population, soil wash, the productivity is stagnating, lowering the labor productivity, increasing the cost of production, adversely affecting the quality and value realization. Due to these adverse soil conditions, poor shade status, lack of cover crops and organic matter, incidence of pests like mites aggravated. Consequent use of higher dose of hard knockdown chemicals repeatedly very often more than the recommended dose led to the resurgence of sucking pests like, thrips and helopeltis pushing up the plant protection cost. Under the prevailing adverse conditions of soil, shade status, Organic carbon level, microbial population, excessive use of chemical fertilizers and herbicides and pesticides there is ample scope for rationalizing the inputs in an ecofriendly integrated way for sustaining the productivity with cleaner produce reducing the cost of production, improving the value realization and economic viability of the plantations.

Factors/aspects for standardization/rationalization:

1. Optimization of soil pH with soil amendments
2. Enriching the Organic carbon
3. Buildup of beneficial microbes
4. Optimize the shade, cover crops and wind barriers
5. Soil conservation through check dams, border grasses, live hedges, temporary multiple shade
6. Liberal use of biofertilizers, biocides encouraging INM, IPM schedules through innovations
7. Effective timely harvest of crop with mechanization and supplemented nourishment
8. Quality up gradation and value addition through diversification and effective marketing

Project Details: The field demo/proving the innovations and package of practices will require at least one pruning cycle/5 years. Comparable sections/divisions of manageable area in a selected garden of the management choice can be taken up during the year 2013. All the inputs and manpower will be provided by the management under the periodical visit and technical guidance of the undersigned. Minimum expected crop increase due to the improved practices and inputs management and guidance is about 30% in the year crop reflecting on the cost of production, improved value realization, reflecting on the economics of the operation. Based on the result, a refined guidelines of SOP will be developed for adoption which will be the property of the management. On acceptance of the proposal, a detailed plan of action along with the budget will be drawn in consultation with the management and work initiated during the year 2013 itself in the nominated garden.

13.6.2 Anticipated benefits at end of the project

Impact of optimization of the basics in fields such as soil Ph, OM status, microbial buildup, shade stand, regulation, etc. on the economy of inputs, cost of production, quality of the produce and value realization, overall economy of the operations thereby the viability of the industry will be known. A nominal saving of fertilizers and plant protection chemicals up to 50% is expected while crop increases up to 35% can be assured during the pruning cycle.

L. Manivel
Consulting Scientist, Plant Physiologist
Coimbatore, Tamil Nadu
India

13.7 A note on glyphosate (C3H8NO5P) on toxicity in plantations and remedial measures

L. Manivel, Plant Physiologist (Retd.), Tocklai, TRA, Jorhat, Assam.

Dear Friends,

It is a pleasure and privilege to stand before you today!

I congratulate the Tocklai, TRA for organizing the 35th conference in a timely manner with this topic of interest. Let us all interact, discuss, and come out with positive feedback and decision to overcome/restore the situation that can be implemented in the fields right away.

We all are aware of the present situation problems of the tea industry in Assam, Doars:

1. Soil and bush health deteriorated
2. Productivity and value realization stagnating
3. COP escalating
4. Excessive, indiscriminate use of chemicals (fertilizers, plant protection, herbicides) aggravated the situation

13.7.1 Remedial/restoration measures suggested

1. The chairman has earlier suggested in ATPA AGM 2017, the reduction of chemicals usage in plantations. Tocklai has already initiated steps for GAP in plantations, and progressing. It is a matter of time for the implementation to get the benefits and turn around. Enough technology and inputs are available for implementation indigenously.
2. Abandon the usage of herbicides in plantation forthwith to detoxify the soil and plants restoring the health and productivity with quality (sustainable). Glyphosate, a weedicide, has already caused enough damage the soil, water, and bush health. The bushes are going through acute stress due to the toxicity of glyphosate build up in soil and plants beyond redemption. It is time to release the plantations from the clutches of the dangerous killer herbicide immediately. Withdrawal of glyphosate itself under the GAP will reflex in 30% of improved productivity by end of the year.
3. The future of the Indian tea industry is good, let us be optimistic and do our best.
4. Problem-oriented research always pays rich dividends.
5. Research projects on the optimization of soil, bush health, and package of practices for ecofriendly cultivation of tea need to be upgraded and fine-tuned for adoption.

L. Manivel

Plant Physiologist (Retd.), Tocklai, Assam, India

Dated, February 22, 2019

13.8 Research work done and contributions made as head plant physiology, Upasi Tri, during 1989–96

Dr. L. Manivel and Team, Dr. S. Marimuthu and Dr. R. Rajkumar

Prelude: I joined the UPASI TRI as **Head plant physiology** with 15 years' experience as a plant physiologist at the premier tea research Institute, Tocklai Experimental Station, Tea Research Association, Jorhat, Assam and 15 years of Research and Teaching in Horticulture, Tamil Nadu Agricultural University, Coimbatore. As a result, the take-off was quick and turnover commensurate with the expectation of the industry, putting the TRI in the limelight.

The state of art of lab facilities with sophisticated instruments was created at UPASI TRI for advanced Physiological/biotechnological research. Tissue culture lab with laminar flow and shade net polyhouse, plant physiology lab with Skalarauto analyzer, HPLC, infrared gas analyzer system, leaf area meter, canopy analyzer, photosynthesis system, plant water status console, phramacia gel electrophoresis system with gel documentation system, ultracentrifuge system, and isotope lab with liquid scintillation counter, etc. have been established availing the central funds and put into use.

Two tea board sponsored projects include

1. Physicochemical studies on Nutrition of tea and
2. role of enzymes in tea manufacture have been successfully completed and the results were published as Research bulletins by the tea board.

Protocol for micro propagation of tea and somatic embryogenesis through tissue culture was developed and the performance of propagules in the field including their variability were assessed availing the Department of Biotechnology grant.

Dr. R. Raj Kumar, Mr. K.S. Murali, Mr. V. Pandidurai have obtained a **PhD in plant physiology** under my guidance with Bharathiar University.

13.8.1 Highlights of research achievements and scientific breakthroughs during the tenure 1989–96

13.8.1.1 Physicochemical studies on nutrition of tea, project sponsored by the tea board of India: principal investigator: Dr. L. Manivel

The long-term field trial was taken up to study the impact of graded dose of nitrogen in the form of urea balanced with Muriate of potash on the bush health, productivity and quality of made tea. The dose of N ranged at 50 kg N/ha, 100 kg N/ha, 200 kg N/ha, 300 kg N/ha, 500 kg N/ha/600 kg N/ha applied in five splits. The cultivars Assam, China, and Cambod were chosen to bring out the genetic potential of cultivar. The trials spread over in five locations in both UPASI as well as member gardens fields in two pruning cycles to bring out the regional influence if any.

Crop was recorded and yield per hectare has been worked out. Soil samples were taken prior to treatment and end of the year, and analyzed for the nutrient profile. Eight year data were pooled, analyzed and the results/inferences are narrated below:

1. Optimum dose of N for sustainable productivity with bush health and quality for south Indian tea plantations is around 300kg N/ha/year applied in four to five splits balanced with K_2O at 1:1 ratio. Out of the 300 kg N broadcast in the form of urea, 10% is lost due to volatilization; 40% is lost due to runoff, leaching losses and fractions get fixed and the balance about only 50% utilized. The percentage of absorption is again governed by the humus content (OM) level in the soil, feeder root proliferation and the soil reaction, shade status, cultural operations, etc. It has been found that even with **150 kg of nitrogen,** yields up to 7000 kg made tea per ha per year is possible with 7% OM, 15,000 population with adequate shade stand, regulated, in clone SA6 under Anamallais conditions.

   It has been proven that nutrition is one of the inputs that determine the productivity. Nevertheless this is not the only factors which determine the yield level. Ultimate productivity is dependent on factors like, cultivar, bush population, soil reaction and topography, organic carbon and humus content, pruning, plucking (timely harvest), number of splits, N:K ratio, and the micronutrient status. Thus, the replacement ratio of 1: 10 is just a guideline but varies with many field situations.

   Increasing the fertilizer dose to compensate the deficiencies in the field is not only uneconomical but also adversely affect the soil, bush health, productivity and quality. One of the perceptible adverse effects of high nitrogen beyond the assimilation capacity (300kg/ha) of the bushes, was a rapid decline in soil pH to hyperacidic levels and depletion of starch reserves, inducing micronutrient Mg, Mn, B deficiencies, and consequently, dieback of bushes and delayed recovery after pruning.

   Assam, China and Cambod, the Chinary hybrids are high input responsive while Assam and Cambod are moderately responsive. Cambod cultivar like CR-6017 being a hybrid between Cambod and China, the clone behaves as the Cambod under favorable conditions while it behaves as Chinary under moisture stress conditions. Cambod clones require more magnesium and sensitive to Mg deficiencies, the annual requirement of zinc and magnesium being 12–15 kg each of zinc sulfate and magnesium sulfate foliar fed at 1% each in the spray fluid at monthly interval.

Extensive field studies of foliar feeding of NK confirmed that urea, MOP cocktail spray of 1% each, not exceeding the total concentration of spray fluid of 3%, given at fortnightly or monthly interval, imparted drought tolerance to the bushes.

The K_2O when foliar fed regulate the gas exchange through stomatal regulation. Thus the bushes go through drought under moderate stresses, preserve the starch reserve and recover early, giving early crop after drought. Added bonus of NK foliar was, that the brightness of the tea liquor was consistently enhanced as revealed by the tasters. Thus NK foliar had been recommended as a practice to overcome the winter drought benefiting the industry.

Isotopic studies on foliar absorption of micronutrient (zinc) studies revealed that almost 50% absorption is completed within 2 h of rain free light period. Further absorption is slow and it takes more than 8 h for 75% absorption. There has been some isotonic effect at 1% fluid concentration there had been always better impact than the higher or lower concentrations.

It has been found that a minimum of 2 h of rainfree period is required for 50% absorption of any foliar and if rain occurs within the 2 h, the spray has to be repeated.

Manivel, L. Marimuthu, S. and Rajkumar, R. Physicochemical studies on Nutrition of tea, Scientific Publication Series No. 9 Tea Board, India, 1997. Page 1–13.

13.8.1.2 *Use of enzymes in tea manufacture (NTRF –tea board project: principal investigator: Dr. S. Marimuthu. Research associate: Dr. M. Gunasekar)*

Coarse plucking especially during monsoon time is common in south India, particularly in the Wyanaad plantations during south west monsoon due to shortage of workers. As a result of delayed/extended round, mature leaves entering the trough causes the Pacha taint and more fibrous tea affecting the value realization. On realizing the chronic serious problem, National

Tea Research Foundation, Tea board sponsored the project to investigate the causes and devise ways and means to overcome the same.

The enzymes found inherently in the tea shoots viz; the Hydrolytic enzymes, cellulose/pectinase, and oxidative enzymes, polyphenol oxidase peroxidases, proteases play vital role in tea manufacture extracting the soluble solids from the cell walls and fibers, contributing for the oxidation (fermentation) of the dhool. The coarse plucked shoots with more mature leaves containing more cholorophyll and carotenoid have been found to contribute for the grassy/rancid flavor with more fibers reducing the cuppage.Usage of hydrolytic enzymes in the tea manufacture was thought of as the solution to this problem. Addition of pectinases and cellulases, were employed in black tea manufacture, with various rates of the enzymes in the CTC rollers as well as in the rollers, fermenting drums/floors.

13.8.2 The results and findings of the investigation are narrated below

1. The excessive embedded fiber and grassy flavor are due to the coarse plucking. The mature leaves and fibers contribute through the breakdown of chlorophyll pigments the grassy flavor (pacha taint).
2. The addition of mixture of cellulose and pectinase enzymes at optimum dose in the rotorvane. CTC third cut stage, or the fermenting floor, imparts the black color of the made tea, improving the cuppage and quality of the brew-brightness.
3. The infusion also becomes bright (golden yellow) improving the value realization.
4. The total solids extracted from the shoots and fibers increased, reducing the embedded fiber while improving the cuppage and value realization.
5. Addition of enzymes in black tea manufacture has become economically viable and the usage has been recommended to the industry with advantage.
6. The usage of Pectinase enzymes has been authenticated by the tea board (commerce ministry) and the concerned section of the FSSAI has been amended for its usage.
7. The results had been published in the form of research bulletin by the tea board and circulated to the industry as the priced publication.

S. Marimuthu and M.Gunasekar, 1998. Tea Manufacture: Role of enzymes. Bulletin published by National Tea Research Foundation. C/o. Tea Board. Page 1–33.

13.8.2.1 DBT project on micropropagation of tea. principal investigator: Dr. L. Manivel, team: research fellows K.S. Murali, & V. Pandidurai

Time lag from release of new clones and the availability of planting material for new expansion is a serious handicap for the industry. Rapid multiplication of the new planting materials through conventional nursery method takes time and paucity of planting materials. With the objective of developing the protocol for micro propagation adopting tissue culture, the DBT project was obtained and the work initiated, employing two research fellows enrolled for their PhD degrees with Bharathiar University Coimbatore. New tissue culture lab was set up and the project was completed within 5 years and the report submitted to the DBT. The two research fellows were awarded their PhD degrees for the research theses.

13.8.3 Salient points of the research findings are narrated below

1. The protocol had been developed using the apical buds as the explants for initiation of callusing, differentiation hardening and establishment
2. The variability of the phenotypes and the establishment in the fields, their performance in regional trials had been evaluated
3. Embroids were developed and attempt was made for breeding for new plant types
4. The protocol for micropropagation has been released to the industry for adoption Vide Planters chronicle, 1996 and subsequently published by DBT in 1999 covering the field performance of TC raised plants
5. A poly house has been erected and employed for screening, hardening and development of plantlets adopting the required nutrition media. Thus a viable **nursery tea special** has been developed suitable to the south Indian tea industry in collaboration with Multiplex of the Karnataka Agro chemicals Company Limited, Bangalore.

S. Marimuthu, 1999. Plant Tissue Culture Technologies for Large scale production and commercialization. Chapter 5. Tea. Published by Department of Biotechnology, Ministry of Science and Technology. Page 45–46.

13.8.3.1 Areas of physiological research remain to be tackled on priority by UPASI TRI

1. Recurring frost in Nilgiris and High Range is a common phenomenon. Effective prophylactic and palliative measures are yet to be worked out based on basic and applied research. This research should be given priority.

2. Optimization of shade in terms of spectrum and mix of species need to be investigated for south Indian tea under the changed climatic conditions, for sustained productivity with consistent quality. Alternate native shade tree species need to evaluated and released to the industry.
3. Buildup of OM and soil amendments for optimizing the soil reaction (pH) to correct the present situation is a priority. Identifying the appropriate inputs, optimize the quantum and method of application, with economics for sustained high productivity need to be investigated and the protocol (SOP) released early for the benefit of the industry.
4. In spite of the high productivity, the tea industry in southern Peninsular becoming uneconomical due to the poor value realization in the market due to unsatisfactory quality. It is high time that all the stake holders join together and correct the situation to meet the standards of Nilgiris logo (GIS) and other tea districts thereby restoring the old glory of the south Indian tea.
5. In order to make the plantation industry viable, mechanization, diversification in to horticultural crops, value addition to tea, adopting eco-friendly cultivation practices, package of viable practices for organic tea in a time bound program is vital

L. Manivel
Consulting Scientist, Plant Physiologist
Coimbatore, Tamil Nadu
India
Dated, December 10, 2018

13.9 Report of visit to Bukhial T. E. by Dr. L. Manivel, Ex-TRA. Plant physiologist

Date of visit: May, 16, 2009.

13.9.1 Introduction

Bukhial TE of McLeod Russel India Ltd., is located in the drought prone Golaghat district with poor distribution of annual rainfall. The soil is of heavy lateritic type which hardens in winter with iron sedimentation/deposit in the subsoil. Elephant herds frequently trespass from the neighboring jungles making the establishment of shade-both temporary and permanent difficult. During the heavy downpour in monsoon, high water table submerging the root zone is common which depletes the starch reserve predisposing the bushes to root diseases and Red spider mites in dry season (winter). The natural outfalls had been blocked by neighbors creating a serious outfall problem to the estate. Consequent to the phased replanting program in vogue, large areas of Assam jat seedlings have been replaced with Tocklai clones which will erode the quality of tea. On the whole, the shade stand is satisfactory. Some of the species have heavy canopy. Incidence of red spider mite and looper attack are evident.

13.9.2 Observations

The visit was mainly concentrated on problematic fields due to time constraints. Soil analysis data and manuring schedules were examined. Spot inspection of the following fields were made and interacted with the Manager and Asst. manager. M5, 4B,3B,B34,B21,19,15, M5,M7,B16,B30.

13.9.3 Comments/suggestions

Soil pH of Sections **B9, M5, M6, M2A** extending to 43.33 ha, was found to be less than 4.0 indicating hyper acidic nature of the soil. Optimum soil pH has to be above 4.5 for economic productivity. Under hyper acidic conditions, feeder roots die, absorption of soil-applied nutrients is hindered and micronutrients (Ca, Mn, B, Mg, Zn) deficiency show up affecting the **productivity and quality**. First split of ground application has been given. There is no point in applying NPK without correcting the soil pH. The soil is moist. It is suggested to apply 2MT of dolomite per hectare and incorporate to correct the ph. The second split may be delayed for a minimum of 5 weeks or compensated along with the third split.

Sec.3B: TV1 planted in 2007 shows severe infestation of red spider mites and loopers. No shade tree has been planted, and no mulching was done either. Centering of the bushes is due. Hard soil pan resulted in twisted root system. Herbicide damage to the collar region is evident.

Suggest: to center the bushes and ring apply decomposed cattle manure or vermicompost liberally, to loosen the soil. Take a furrow in alternate rows and sow Tephrosia (Bogamedula) or sunn hemp (Crotaleria) as cover crop, establishing temporary shade Indigofera or Bokane. The bushes need to be mulched.

B34: Localized high water table. Drainage has since been improved and bushes are recovering. Suggest to apply decomposed cattle manure (5MT/ha) with Trichoderma to boost the growth and prevent the root rot diseases.

B21, 19,15,M5, M7, B16: A trough with high water table leading to dieback and canker.

Suggest HRP to repair the frame, restoring the sap flow with soil leveling and amendments. **High water table in the root zone during monsoon due to outfall problem is the principal cause for dieback and decline.**

Deep diversion drain is being taken. In order to prevent landslide, the borders may be planted with vettiver (khus khus) grass.

To reduce the water table below the root zone an attempt may be made to plant out the following fast growing trees species in the peripheries of the trough/hullah:

1. Albizzia mollucana
2. Eucalyptus grandis
3. Acacia planiferons
4. Cassia siamea
5. Spathodia
6. Erythrina (dadap)
7. Glyricidia
8. Melia azadirach (Bokane)
9. Grevillea (silveroak)
10. Kadam
11. Karanji (Pongamia)
12. Ajhar (Lagastroemia) Jharul

Rainwater Harvest: As the outfall problem is serious, the trough in this area may be utilized as tank to conserve the water during monsoon, which can be utilized during winter for irrigation. Since the soil is high in iron, the rainwater harvested and conserved can be utilized.

M7: The P126 is a sensitive quality Chinary clone. Due to prolonged high water table in the root zone, depleting the root reserve, the bushes are declining in patches while the hardy infills are doing well. The recovery is poor and dieback is evident causing vacancies.

Suggest to uproot the area, rehabilitate and plant out with hardy quality Biclonal seed stocks like TS 569(Heleaka22/14xTv20), TS 491(Tv1xS3A1) TS464 (TV1xTV20).

B30: Patches under the shade trees have not recovered after pruning. Black rot (*Corticium theae*) has affected the bushes defoliating under heavy shade of the tree canopy. Under heavy soil and high water table root diseases like violet, pink, brown, and charcoal stem rot have aggravated the problem. The frame is also affected by poria (branch canker) by physical injuries by hail or goat damage.

Suggest to regulate and open up the shade trees by pollarding before monsoon and take effective measures against black rot. Hoe the soil and apply OM + Trichoderma (5kg/ha)+ pseudomonas (20kg/ha) against root diseases by season end. Give a spray of Trichoderma + pseudomonas 2 kg each per drum to the frame at the time of prune.

13.9.4 Productivity and quality

The crop of the garden is said to be stagnating and of late erosion in quality. The following factors may be responsible for the same:

13.9.4.1 Productivity

1. High water table depleting the root reserve weakened the bushes, which become vulnerable to drought and mite incidence, leading to defoliation and dieback thereby delayed recovery affecting the early crop.
2. Heavy soil and poor drainage resulted in shallow root system, poor root reserve, predisposed to the incidence of root diseases resulting in poor assimilation and response to NPK applied.
3. Heavy canopy of the shade species contributed to the incidence of black rot leading to the defoliation and delayed recovery of bushes under the shade trees, affecting the early and total crop.
4. Due to repeated chronic water logging in the root zone and drought, the bushes are weakened and branch canker set in, in many low lying fields particularly so in Tv1 and P126 clones. This may be due to sun scorch, injury by hail and goat damage, affecting the sap flow resulting in poor response to inputs.
5. Being drought prone area, incidence of mites and sucking insects is severe. Of late incidence of loopers also affects the crop, escalating the cost of production.
6. Use of herbicides in young tea also promotes canker incidence resulting in delayed ground cover and frame formation which affect the early crop and productivity.
7. The list of approved plant protection chemicals deserve a second look and reconsideration. Decis (delta methrin) has a history contributing resurgence of sucking insects like thrips, greenflies, helopeltis and loopers. Any repeat application of decis is expected to have adverse effect on effective pest management. Alternatives deserve to be considered and IPM enforced.
8. Sections yielding below garden average due to various constraints should be medium pruned, soils amended, restoring the health and productivity.

9. **B**ukhial clonal blend indicates

TV1	30%
TV17	15%
Tinali17	20%
S3A3	20%
T3 E3	15%

The clonal blend is highly biased toward quality, takes longer time for peak productivity (except Teen Ali). Inclusion of Tocklai seed stocks like TS 464,491,589 worth to be considered in the replanting program.

13.9.4.2 Quality

1. Replanting program is in vogue and the quality seed jats are being replaced with clones, **which will have an impact on the quality**.
2. Soils of many sections are hyper acidic in which essential nutrients, macro and micro, are not readily available which will affect the quality of the produce.
3. Consequent to the outbreak of pests like mites, loopers and helopeltis, usage of knockdown chemicals increased. Repeated use of any chemical, that too higher than the recommended dose at frequent intervals, is bound to affect the quality, leading to the resurgence of new pests. IPM schedule as recommended by Tocklai should be followed for preserving the balance of predators to pests.
4. The quality of bore well water used for winter irrigation should be checked. High iron or oil smear in irrigation water is expected to affect the quality of the produce. Rainwater harvest is in focus to mitigate the situation.
5. Excessive use of herbicides, reduction in K2O application, low pH of the soil are a few factors likely to have impact on the quality of the produce.

I take this opportunity to convey my heartfelt thanks to the Manager Mr. Sandeep Nagalia for the excellent hospitality, sparing his valuable time to show me round the fields providing the required information. I am also grateful to the management for permitting me to visit the garden and interact with the Managers.

L. Manivel
Plantations Consultant

13.10 Decline of Tuckda 383-Long View, Darjeeling

Date of visit: May 10, 2009

Long View is located at the foot of Darjeeling hills. The clone Tuckdah 383 is a Chinary hybrid with yield potential up to 16 quintals.

Present: Mr. R.S Tiwari (Supdt), Mgrs:Guldip, Manoj, and Rajinder

13.10.1 Observations

The bushes are going through the stress leading to dieback and defoliation. Recovery is poor after the stress. The iodine test indicated the poor starch reserve status in roots confirming the weakness of the bushes. Stray incidence of red rust was evident indicating the high water table (localized water logging) during monsoon as well as inadequate potash nutrition. The shade stand in the field also desired to be improved. The dieback from the tophamper has been confirmed due to the new strain of red rust, which is not visible to the naked eye. As the algal parasite blocks the sap flow and spreads through plucking wounds harboring the crow's feet the disease has to be tackled judiciously, adopting cultural, chemical (Bordeaux mix; copper oxychloride), nutritional, and field sanitation and bush sanitation measures. It is more of a **physiological disorder** which can be corrected and yield improved with in a pruning cycle.

13.10.2 Suggestions

1. Apply two rounds COC spray to control the algal parasite after level skiffing the dieback
2. A round of foliar with MOP+MgSo4 +Boric acid to correct the nutritional disorders
3. Follow it up with a round of biocides like Trichoderma + Pseudomonas + Verticillium@5kg each per hectare along with molasses or jaggery
4. Establish a new canopy by liberal plucking or addition of a new leaf
5. Give extra dose of MOP during ground application.
6. Deepen the peripheral drains, to remove the localized water logging(-high water table during monsoon)
7. Give a round of foliar of biocides against Tea mosquito with Beauveria +Metarrhizium with molasses or cane juice alternated with herbal extracts
8. Adopt 3 year pruning cycle. Rest the bushes at least for a month, delay the pruning to Jan, and leave a breather(lung) as an insurance and quick recovery

13.10.3 Diagnosis and recommendations

It is a physiological disorder specific to the clone which can be corrected with combinations of cultural and management practices. The bushes need not be uprooted. The bushes can be inspected again in October to review the progress and response of the bushes. If necessary an in house training program for the executives can be organized during the winter.

L. Manivel

13.11 Salient points on young tea management, Meisamari, Nigeria

1. Cultivar: China × Cambod hybrid clones or seedlings, graft, cuttings, micropropagated for uniformity
2. Population:15,000/ha., staggered double hedge, contour planting
3. Spacing, pits etc: 60 × 60 × 75 cm, pit size.45 cm × 45 cm minimum.
4. Planting: inputs per pit, Rockphos 100–150 g, Decomposed Cattle manure, compost, 2–3 kg, Phosphobacter, Metarrhizium, Paecilomyces 5 g each, mix with top soil and fill up to the brim, take a hole and plant in the middle of pit. Ramp the soil, irrigate, and stake against wind or use thatch cover.
5. Aftercare: Mulch around the collar avoiding touch to the central stem/trunk. Cover crops like Crotalaria sown, and Irrigated, herbal weeds preserved Plant temporary, permanent shade trees, horizontal furrow drains, staggered and later filled up with dried leaves, trash, and weeds for conserving soil moisture.
6. After care:
   a. Spray Goal-Oxyfluorfen around the collar within a month of planting to induce laterals, as a preemergent herbicide
   b. Center the bushes when the lateral buds sprout and spread, cut the main stem 12–15 cm height, with three to five lateral sprouts
   c. Give a foliar of micronutrients and Triacontanol, Humic acid and Gibberellins, no herbicides in YT
   d. Tip the laterals 35 cm and form table for quick ground cover suppressing the weed growth

e. Step up/raise the table @ 5 cm per year to form the table through formative prune in third or fourth year depending on growth, clones. Open the center, de-snag, fill up the pockets and form the table frame for future pruning and harvesting.

f. The YT management lays the foundation for the future productivity, quality, bush Architecture and Longevity/life span of tea plantations, Adequate care pays in the long run

L. Manivel
Plantation Consultant

References

Barua, D.N., 1989. Science and Practices in Tea Culture. TRA, Kolkata, India, pp. 45–153.
Baruah, P., 2006. The Tea Industry of Assam-the Origin and Development. EBS, Guwahati, India, p. 326.
Baruah, P., 2019a. Tea. An important constituent of biodiversity of North East India. Int. J. Agric. Sci. 19–33.
Baruah, P., 2019b. Tea Industry of Kenya, Its Assam Linkage, Purple Tea and Potential in Assam.
Carr, M.K.V., 2018. Advances in Tea Agronomy. Cambridge Univ. Press, Uk London, p. 456.
Chatterjee, R.N., 2018. Management of Water and Fertilizer (Plant Nutrients) for Sustainability in Tea. ATPA Yearbook, pp. 35–36.
Kempf, J., 2019. Foliar Application on Photosynthesis-Organic Foliar Applications Accelerate Growth.
Owuor, P.O., 2001. Effects of fertilizers on tea yield and quality. A review with special reference to Africa and Sri Lanka. Int. J. Tea Sci. 1 (1), 1–11.
Sarma, P.K., 2018. Water Related Hazards Faced by Tea Estates of Assam and Need for Appropriate Sustainable Management ATPA Year Book, pp. 37–41.
Tocklai, 1988. Hand Book on Tea, 2012 Tocklai Tea Encyclopaedia.

Further reading

Chelvan, T., 2015. Personal Communication With Dr. Thamizhchelvan. Vishakan Biotech, Pvt. Ltd., Coimbatore.
Durairaj, j, Radhakrishnan, B., Hudson, J.B., Muralidharan, N., 2019. Guidelines on Tea Culture in South India. UPASI, Coonoor.
Dutta, A.K., 2018. The Brahmaputra-Arbiter of Assam's Destiny, pp. 15–18.
Goswami, D., 1992. A Study of the Tea Labor Population of Assam, Brahmaputra Valley of Assam (Ph.D. thesis) submitted to Dept of Economics, Guwahati University, Giuwahati.
Griffiths, P., 2019. Tea Industry Problems. Planters Chronicle, pp. 33–42. From the History of Planter's Chronicle, Jun 1, 1952.
Kannan, S., 2010. Foliar application for sustainable crop production, Chap. 10. In: Genetic Engineering, Biofertilization, Soil Quality and Organic Farming. Springer, pp. 371–402.
Lin, M.-L., 2015. Technical Development of Production and Products in Organic Farming. IJTS11 (3&4), pp. 23–32.
Manivel, L., 1967. Effect of certain fertilizer treatments on the fruitfulness of Grape buds. In: Anabe Shahi, Variety.
Manivel, L., 1973. Influence of Growth Regulators on Photosynthesis, Translocation and Stomatal Activity of *Vitis vinifera*, L.1973. Dissertation Submitted as Partial Satisfaction of the Requirements for the Degree of Doctor of Philosophy in Plant Physiology, in the Graduate Division of, UCD, California, USA.
Manivel, L., 1998. Tea: botany and horticulture, chap. 7. In: Janick, J. (Ed.), Horticulture Reviews, vol. 22, pp. 267–295. USA.
Manivel, L., 1999. Physiology of tea productivity, pp. 463–480. In: Jain, N.K. (Ed.), Global Advances in Tea Science. Aravalli Books International, Pvt. Ltd., New Delhi.

Manivel, L., Hussain, S., 1982. Photosynthesis in tea. 1. Contribution of photosynthates to pluckable shoots by maintenance leaves. Two Bud 29, 13–16.

Manivel, L., Hussain, S., 1986a. Photosynthesis in tea. I&II. Contribution and direction of movement of photosynthates from leaves vs crop distribution. Two Bud 29, 13–16: 33:90–93.

Manivel, L., Hussain, S., 1986b. Relative Sink capacity of developing shoots. Two Bud 33, 90–93.

Manivel, L., Hussain, S., Sarma, A.K., 1981. Seasnal changes in distribution of photsynthates to pluckable shoots by maintenance leaves. Two Bud 29, 13–16.

Muralidharan, N., 1991. Pest Management in Tea. Published by Director. UPASI Tea Research Institute, Valparai.

Nakagawa, M., 2015. In: Apostolides, Z. (Ed.), The History of Tea Components, p. 279.

People Who Have Left an Indelible Mark on Assam Tea, 2018. Image, ATPA Year Book. pp. 55. (permission obtained).

Phukan, I.K., Bhagat, R.M., Dutt, J., 2015. Response of tea to micronutrients application on its yield and quality. Int. J. Tea Sci. 39–47.

Plantation Communities; a Stock Taking 2019–2020 Planter's Chronicle, August 2021, pp. 1–12.

Prahat Bezborurah, 2018. Tea Blues. The Chairman's Address, Tea Board, India, ATPA Year Book, pp. 12–13.

Rajkumar, R., Manivel, L., Marimuthu, S., 1998. Longevity and factors influencing photosynthesis in tea leaves. Photosynthetica 35, 41–46.

Ramasamy, V., 2019. The Tea Tasting. Planters Chronicle.

Robert, G.R., Keys, A j, 1978. The Mechanism of Photosynthesis in tea plant (*Camellia sinensis*, L). J. Exp. Bot. 29, 1403–1407.

Sharma, V.S., 2011. A manual of tea cultivation. Int. J. Tea Sci. 1–148.

Sharma, V.S., Kumudini, G. (Eds.), 2018. Global Tea Science- Current Status and Future Needs. MTBurleigh Dodds Science Publishing, Ltd., No. 41, pp. 1–531.

Sidhu, H., 2019. Integrated weed management. The Tea Times, Kolkota.

Singh, G.N., Manive, L., 2012. Banjhi Management in Tea (Tea Times, Assam News and Reports, Kolkata, India).

Tea Growers Hand Book 5th Edition Tea Research Foundation, Kericho, Kenya.

Tea Times, 2005. In: International Tea conf., New Delhi 2005. Welfare Measures for Tea Plantation Workers.

The Soils, Weekly News, August 2019. Foundation of Natural Farming, USA.

Thomas, J., 2019. Annual Report, Statistics. On Crop Production, Export Price, 20 Kolkata, India. Excerpts Incorporated, Permission Obtained.

Willis, H., 2019. The Soils Chap. 3, Weekly News, Sept., 2019, Book of the Week, The Weekly Feature of Extension Acres, USA.

Index

'*Note*: Page numbers followed by "f " indicate figures and "t" indicate tables.'

R

S

T

U

V

W

Kolibri 360 GmbH
Ferdinand-Jühlke-Straße 7
99095 Erfurt, Deutschland
produktsicherheit@kolibri360.de